T0306001

Climate Change and Carbon Recycling

Climate Change and Carbon Recycling: Surface Chemistry Applications describes the application of surface chemistry methods for carbon capture and recycling in relation to climate change and atmospheric CO_2 levels. The text is suitable for online education, with both basic and educational descriptions of the climate change process and carbon recycling methods like the adsorption and absorption of CO_2 on solids. This book leads to a better understanding of a complex phenomenon and highlights the importance of CO_2 capture and sequestration for the future to enable the utilization of fossil fuels without contributing to atmospheric greenhouse gases.

Features

This unique volume specifically:
- Highlights the surface chemistry aspects of carbon capture and recycling (CCR).
- Fills the need for an online textbook edition, which provides a basic and educational description of the climate change process and CCR.
- Describes the application of surface chemistry methods for CCR, such as adsorption/absorption of CO_2.
- Discusses the importance of recycling in reducing and controlling the concentration of carbon dioxide in the air.
- Describes the importance of the technology related to CCR and carbon capture sequestration (CCS) from fossil fuel energy plants as a means of CO_2 control.

Sustainability: Contributions through Science and Technology

Series Editors: Thomas P. Umile, Ph.D, Villanova University, Pennsylvania, USA
William M. Nelson, US Army ERDC, USA

Preface to the Series

Sustainability is rapidly moving from the wings to center stage. Overconsumption of non-renewable and renewable resources, as well as the concomitant production of waste has brought the world to a crossroads. Green chemistry, along with other green sciences technologies, must play a leading role in bringing about a sustainable society. The Sustainability: Contributions through Science and Technology series focuses on the role science can play in developing technologies that lessen our environmental impact. This highly interdisciplinary series discusses significant and timely topics ranging from energy research to the implementation of sustainable technologies. Our intention is for scientists from a variety of disciplines to provide contributions that recognize how the development of green technologies affects the triple bottom line (society, economic, and environment). The series will be of interest to academics, researchers, professionals, business leaders, policy makers, and students, as well as individuals who want to know the basics of the science and technology of sustainability.

Michael C. Cann

Environmentally Friendly Syntheses Using Ionic Liquids
Edited by Jairton Dupont, Toshiyuki Itoh, Pedro Lozano, Sanjay V. Malhotra, 2015

Catalysis for Sustainability: Goals, Challenges, and Impacts
Edited by Thomas P. Umile, 2015

Nanocellulose and Sustainability: Production, Properties, Applications, and Case Studies
Edited by Koon-Yang Lee, 2017

Sustainability of Biomass through Bio-based Chemistry
Edited by Valentin Popa, 2021

Nanotechnologies in Green Chemistry and Environmental Sustainability, 2022

Towards Sustainability in the Wine Industry by Valorization of Waste Products: Bioactive Extracts
Edited by Patricia Joyce Pamela Zorro Mateus and Siby Inés Garcés Polo, 2023

Climate Change and Carbon Recycling: Surface Chemistry Applications
K.S. Birdi

Climate Change and Carbon Recycling
Surface Chemistry Applications

K. S. Birdi
Consultant, Holte, Denmark

CRC Press
Taylor & Francis Group
Boca Raton London New York

CRC Press is an imprint of the
Taylor & Francis Group, an **informa** business

First edition published 2024
by CRC Press
2385 Executive Center Drive, Suite 320, Boca Raton, FL 33431

and by CRC Press
4 Park Square, Milton Park, Abingdon, Oxon, OX14 4RN

CRC Press is an imprint of Taylor & Francis Group, LLC

Library of Congress Cataloging-in-Publication Data
Names: Birdi, K. S., 1934- author.
Title: Climate change and carbon recycling : surface chemistry applications / authored by: K.S. Birdi.
Description: First edition. | Boca Raton : CRC Press, 2024. | Series: Sustainability : contributions through science and technology | Includes bibliographical references. | Summary: "Climate Change and Carbon Recycling: Surface Chemistry Applications describes the application of surface chemistry methods for carbon capture and recycling in relation to climate change and atmospheric CO_2 levels. The text is suitable for online education, with both basic and educational descriptions of the climate change process and carbon recycling methods like the adsorption and absorption of CO_2 on solids. This book leads to a better understanding of a complex phenomenon and highlight the importance of CO_2 capture and sequestration for the future to enable the utilization of fossil fuels without contributing to atmospheric greenhouse gases"—Provided by publisher.
Identifiers: LCCN 2023030263 (print) | LCCN 2023030264 (ebook) |
ISBN 9781032291543 (hardback) | ISBN 9781032291550 (paperback) |
ISBN 9781003300250 (ebook)
Subjects: LCSH: Carbon dioxide mitigation. | Carbon sequestration. |
Climate change mitigation. | Surface chemistry
Classification: LCC TD885.5.C3 B56 2024 (print) | LCC TD885.5.C3 (ebook) |
DDC 660/.293—dc23/eng/20231018
LC record available at https://lccn.loc.gov/2023030263
LC ebook record available at https://lccn.loc.gov/2023030264

ISBN: 978-1-032-29154-3 (hbk)
ISBN: 978-1-032-29155-0 (pbk)
ISBN: 978-1-003-30025-0 (ebk)

DOI: 10.1201/9781003300250

Typeset in Times
by codeMantra

Contents

Preface

The environment of the earth comprises different kinds of surface chemistry aspects. This characteristic has been recognized in the current literature. The chemical evolutionary equilibrium (CEE) has thus been the basis of the living species on the earth. The environment has thus stabilized, in spite of these different interfaces. The current subject of climate changes, i.e., change in the surface temperature of the earth, is considered. The latter phenomenon is related to CEE and interfaces. The phenomena of CEE were based on literature data (Melvin Calvin, 1969: Nobel Prize 1961; private communications). The application of carbon capture recycling and storage (CCRS) technology is delineated. During the evolution (over 4 billion years), the environment has achieved pseudo-equilibrium. These parameters have adjusted to changing environments (e.g., climate, carbon emissions from fossil fuels, earthquake, and wars). Mankind is the only living species, which has interacted with the environment. This has been very obvious, especially after the industrial revolution (about 200 years ago). This book delineates the surface chemical aspects of climate and its relationship to CEE. The surface temperature of the earth is reported to have increased by about one degree, since the Industrial Revolution (Bill Gates, 2021). The manuscript highlights the subject through suitable examples. These examples are intended to highlight the significant items. A short, but essential, description is included, as regards the surface chemistry principles. The aim of this is to help the reader in following the text seamlessly. The subject matter is therefore arranged in a useful manner, with the intention of smooth understanding of the CCRS technology. More advanced analysis is separately described in Appendices. This is intended for those readers, who wish to follow the subject matter to a higher level.

Author

K.S. Birdi studied at (B.Sc. (Hons)) Delhi University, Delhi, India, and in 1952, he then traveled to the United States for further studies, majoring in chemistry at the University of California at Berkeley. After graduation in 1957, he joined Standard Oil of California, Richmond.

Dr. Birdi moved to Copenhagen in 1959, where he joined Lever Bros., as Chief-Chemist, Development Laboratory. During this period, he became interested in surface chemistry and joined, as assistant professor, in the Institute of Physical Chemistry (founder of the institute: Professor J. Bronsted), Danish Technical University, Lyngby, Denmark, in 1966. He initially did research on surface science aspects (e.g. detergents; micelle formation; adsorption; biophysics). During the early exploration and discovery stages of oil and gas in the North Sea, he got involved in Research Science Foundation programs, with other Research Institutes around Copenhagen, in the oil recovery phenomena and surface science. Later, research grants on the same subject were awarded from the European Union projects. These projects also involved extensive visits to other universities and an exchange of guests from all over the world. Dr. Birdi was appointed as Research Professor in 1985 (Nordic Science Foundation) and was then appointed, in 1990 (retired in 1999), in the School of Pharmacy, Copenhagen, as professor in Physical Chemistry.

There was continuous involvement with various industrial contract research programs throughout these years. These projects have actually been a very important source of information in keeping up with real-world problems and helped in the guidance of research planning at all levels.

1 Carbon Capture-Recycling/Surface Chemistry Aspects

1.1 INTRODUCTION

The subject matter in this manuscript relates to man-made industrial life essential activities and its relative effect on *climate change* (e.g. temperature and pollution (e.g. air-pollution; drinking-water pollution; ocean/river/lake pollution). Especially after the industrial revolution, the combustion of *fossil fuels* such as coal/oil/gas has been increasing, as well as the population. The increasing industrialization and global activity have further added to attempts to affect the environmental state of earth (Birdi, 2020; Gates, 2021; Lomborg, 2007, 2022; Kemp et al., 2022; Rosenzweig et al., 2021; IPCC, 2011; Epstein, 2022) (Figure 1.1).

The mankind has evolved on the earth over a few hundred-thousands of years (Calvin, 1969). The phenomena of fire, was invented by man a few thousands of year ago.

Later, about 200 years ago, crude oil and gas reservoirs were discovered. This led to innovations and gave rise to the so-called industrial revolution. These innovations developed continuously. Currently, the usage of fossil fuels (e.g. wood + crude oil + natural gas) has a significant effect on the environment. This is related to the daily fossil fuel application in the maintenance of life necessities (Figure 1.2).

The sun radiation-atmosphere and atmosphere-earth systems thus give rise to two different phases interacting at an interface. This phenomenon gives rise to surface chemistry aspects.

One of the main products produced by the **combustion** of fossil fuels (used to produce energy; food; transport; human health; household; infrastructure) has been gaseous carbon dioxide (CO_{2gas}).

EXAMPLE

<Fossil Fuels – Combustion -> Energy

<Combustion -> Addition of Carbon Dioxide to Atmosphere

However, the phenomena of *photosynthesis* existed prior to living species (Calvin, 1969). Latter is based on:

<sunshine + carbon dioxide, CO_{2gas} + water

Thus, fire invention added extra CO_{2gas} to the atmosphere. GHG.

DOI: 10.1201/9781003300250-1

EXAMPLE

<EVOLUTIONARY PHOTOSYNTHESIS

    <1>Sun radiation
    <2>Carbon Dioxide Gas
    <3>Water
    <4>Chemical Reaction

$$6\ CO_{2gas} + 6\ H_2O \rightarrow (\text{Sun radiation}) \rightarrow C_6H_{12}O_6 + 6\ O_2$$

This man-made phenomenon mainly takes place only on 30% of earth surface.

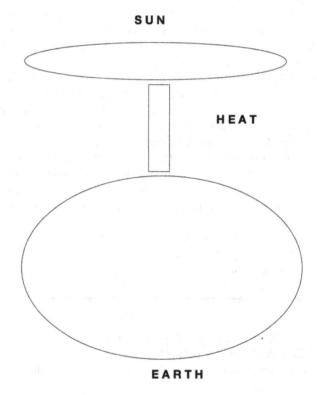

FIGURE 1.1 Earth receives heat radiation from sun [5,000°C] – atmosphere – earth [−50°C to +50°C] (surface (average) temperature of earth).

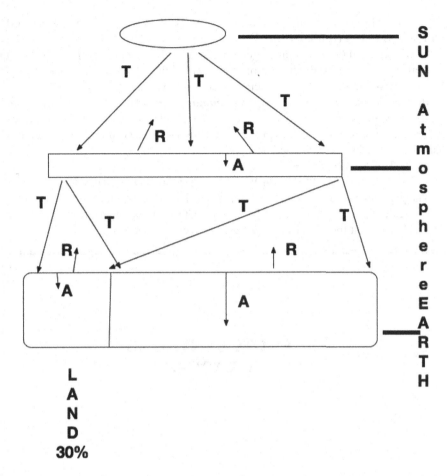

FIGURE 1.2 Heat (radiation) balance from sun to earth surface (30 % land + antarctica) + 70 % oceans (plus: lakes + rivers) (arrows signify: (T)transmit/(R)reflect/(A)absorb).)

EXAMPLE

        <MAN-MADE PRODUCTION OF CARBON DIOXIDE (CO_{2gas})
FROM
        THE COMBUSTION OF FOSSIL FUELS>

<Fossil fuels + oxygen (atmosphere) -> $CO_{2,gas} + H_2O$

    <ADDITION OF MAN-MADE CARBON DIOXIDE ($CO_{2,GAS}$) TO
ATMOSPHERE (AT EQUILIBRIUM)
    <INCREASING *EQUILIBRIUM CONCENTRATION OF CARBON DIOXIDE
IN ATMOSPHERE* (CURRENTLY 420 PPM) AND ITS EFFECT ON CLIMATE
CHANGE (HIGHER TEMPERATURES) DUE TO ITS GREENHOUSE EFFECT

(P.S. worldwide crude oil is used, in the past few decades, at a rate of about 100 million barrels/day (Birdi, 2020))

However, fossil (fuel) substances (comprising *wood; coal; crude oil; natural gas*) have also been used for production of other commodities for mankind than energy. These are essentially medical/medicines/vaccines/food-packaging-transport/etc.

Besides, it is known that the process of ***photosynthesis*** has existed on earth from pre-living species era. This evidences the presence of carbon dioxide in the atmosphere for billions of years (Calvin, 1969). The photosynthesis process (plants/food/fisheries) is fundamentally based on:

>PHOTOSYNTHESIS ==

> SUN SHINE + WATER + CARBON DIOXIDE (CO_2)

The phenomenon of photosynthesis on the earth requires <sunshine + water + carbon dioxide ($CO_{2,gas}$). This means that there is CO_{2gas} in the atmosphere. The latter thus indicates that there is also the GHG phenomenon (Figure 1.3).

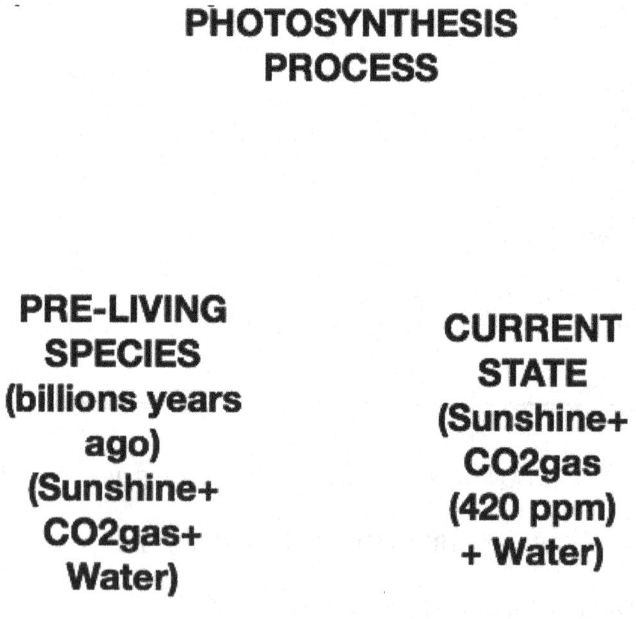

PHOTOSYNTHESIS PROCESS

PRE-LIVING SPECIES (billions years ago) (Sunshine+ CO2gas+ Water)

CURRENT STATE (Sunshine+ CO2gas (420 ppm) + Water)

FIGURE 1.3 Evolutionary aspects of photosynthesis (and GHG) (billions of years ago) from pre-living species to current stage on earth.

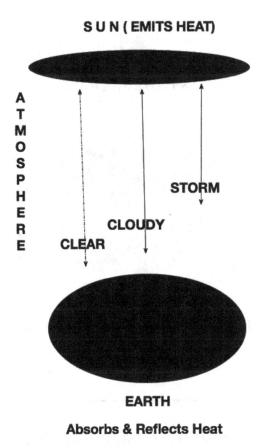

FIGURE 1.4 Sun (heat) – atmosphere – earth system.

With regard to the environment, the current trend of *increasing concentration of carbon* in air has been related to its effect on the temperature of global climate (Birdi, 2020; Gates, 2021; Kemp et al., 2022; IPCC, 2011; Rosenzweig et al., 2021). However, carbon dioxide, CO_{2gas}, has existed since photosynthesis began. In other words, the GHG effect has changed during this geological era.

However, besides this parameter, one also needs to consider other aspects of the climate dependency on man-made activities on earth (Figure 1.4).

The essential role of the atmosphere and its contribution to the chemical evolution on the earth has been recognized in literature (Calvin, 1969; Birdi, 2020; Kemp et al., 2022).

Furthermore, it is also recognized that in all natural and man-made phenomena, **surface chemistry** principles have played an important role.

This manuscript addresses the *surface chemistry principles* which have been recognized to assist in the control/mitigation of this man-made process (Chattoraj and Birdi, 1984; Birdi, 2020; Kemp et al., 2022).

The different interfaces are recognized and analyzed (Figure 1.5).

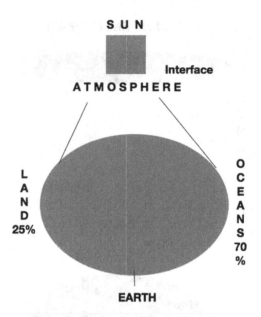

FIGURE 1.5 Sun – Interface – Atmosphere; Atmosphere – Interface – Land; Atmosphere – Interface – Oceans.

EXAMPLE

>CARBON RECYCLING/SURFACE CHEMISTRY ASPECTS
>CARBON RECYCLING AND CLIMATE CHANGE
>CARBON   RECYCLING/MAN-
MADE FOSSIL FUEL COMBUSTION (ADDITION OF CARBON DIOXIDE TO ATMOSPHERE)

This is the primary aim of the present manuscript and to add most updates on this subject. The applications of essential surface chemistry aspects are pertinent and delineated accordingly.

EXAMPLE

<CLIMATE AND SURFACE CHEMISTRY >
<SUN (HEAT EMISSION) – ATMOSPHERE:::: *INTERFACE*
<ATMOSPHERE   –   EARTH   (LAND/OCEANS/LAKES/
RIVERS/ARCTICA)

<GAS (ATMOSPHERE) – LIQUID (OCEANS/LAKES/RIVERS)::: *INTERFACE*
    <SOLID (LAND/ANTARCTIC/SNOW) – ATMOSPHERE::: *INTERFACE*

It is found that these latter quantities are interrelated. There is thus a need to mitigate the effect of increasing usage of fossil fuels and increasing negative effect on the climate and environment, in order to mitigate this phenomenon.

EXAMPLE

>CLIMATE-FOSSIL FUEL-CARBON DIOXIDE (CO_{2GAS})
    >MITIGATION OF CARBON (*CAPTURE-RECYCLING-STORAGE*)
>CARBON CAPTURE RECYCLING AND STORAGE (**CCRS**)

However, in addition, one finds that the man-made increasing degree of pollution (general), particularly related to fossil fuels usage (direct or indirect) needs to be mitigated. The CCRS technology is being applied to many different fossil fuel energy plants (IPCC, 2011; Birdi, 2020). The term <pollution> applies to all

Man-made Activities:
Fossil Fuel Combustion +
Transport +
Energy Production +
Medical + Food / Fisheries

Air Pollution
Oceans Pollution
Rivers/Lakes Pollution
Drinking Water Pollution
Noise Pollution

FIGURE 1.6 Man-made activities and pollution of earth environment.

man-made activities which interferes with the normal environmental phenomena (Figure 1.6).

It is also of interest to consider the *chemical evolutionary equilibrium* (CEE) mass balance of this subject. The CEE is related to the *interfaces* found in the sun – atmosphere – earth system. The CEE analysis also indicates that these phenomena have changed over this geological era. Thus, the climate models also estimate that these will continue in near future.

EXAMPLE

<CURRENT: SUN – ATMOSPHERE – EARTH>
  <Amount of $CO_{2,gas}$ in atmosphere = 750 Gtons>
  <Amount pf $CO_{2,aq}$ in oceans/lakes/rivers = >50,000 Gtons.>
  <Amount of $CO_{2\ bound}$ through photosynthesis in plants = 450 Gtons
  (Gton = billion ton = 10^9 Tons)
  <CREATION OF EARTH (4–5 billion years ago) = Very hot planet = No atmosphere = No photosynthesis = No liquid water = No living species>
No-GHG

In order to discuss the present state of carbon dioxide, CO_{2gas}, it is useful to consider the creation of earth.

In other words, the composition of the atmosphere was different in the beginning, i.e., prior to oceans/lakes/rivers. As the planet earth cooled, mostly on the surface, and a solid crust was formed (or is partially (changing) forming whenever there is earthquake eruption ( ).

1.2   GREENHOUSE GAS (GHG) EFFECT AND SOLAR SYSTEM

The most unique characteristic of the sun – earth system is that the latter is surrounded by a layer of gas (atmosphere), which consists of different gases (e.g. nitrogen, oxygen, carbon dioxide, water vapor, and traces of other gases). The gas molecules, as found in the atmosphere, are known to be attracted to the earth by gravity forces (Fowler, 1967; Clifton, 2017). The composition of the atmosphere has reached an (pseudo) equilibrium over the billions of years. Some of these atmospheric gases are known to absorb infra-red spectra and thus they reflect heat waves. These gases are termed as *the greenhouse gas*, GHG. Some of these gas molecules are given as follows:

- carbon dioxide (CO_{2gas})
- methane gas (CH_4)
- NO_2
- water vapor (only reflection)

This occurs simultaneously to other factors, such as release of GHG (such as carbon dioxide; methane) (Birdi, 2020) in the mitigation process; important surface chemistry principles are present. The latter is described as needed in the context. However, none of these phenomena are evenly situated over the surface of the earth. This means that these phenomena vary from time and place.

The current relationship between the effect of increasing GHG by man-made technical activities is reported to have been the cause in the semi-quantitative temperature measurement (IPCC, 1995, 2011; Birdi, 2020; Gates, 2022). The effect of different GHGs is found to be different (Appendix A).

EXAMPLE

<GHG EFFICIENCY OF DIFFERENT (MAN-MADE) POLLUTION RELATED GASES>

NITROUS OXIDE (N_2O_{gas}) >>>> METHANE (NATURAL GAS: CH_{4gas}) > CARBON DIOXIDE: CO_{2gas}

OTHER (WATER VAPOR: CLOUDS).....................................(ATMOSPHERE: VARIABLE)

It is important to mention that the GHG effect by CO_{2gas} is only observed in its gaseous state.

Further, the gas density decreases with height from earth. In fact, the gas density is almost absent at a height of 20 km. This means that any GHG effect is varying with height; and it is nonlinear. Additionally, since the composition of GHG gases is different at different time/place (such as over oceans), the GHG effect will also be erratic. In other words, the degree of GHG through the billions of years has changed as the composition of GHG gases has varied.

It is thus seen that the GHG effect on climate is a cumulative effect and has been considered to be much more complex and needs more concerted analyses (Salty, 2012; Kemp et al., 2022; Birdi, 2020; Lomborg, 2022; IPCC, 2019, 2011).

Furthermore, the sun – atmosphere – earth constellation in this phenomenon is known to be of main interest. In addition, the evolutionary aspects of living species (especially human kind) are mentioned, (briefly) wherever it seems to be relevant (Figure 1.7).

However, it is thus important to briefly describe the planet earth and its existence in the solar system.

In the solar system, planets revolve around the sun following a given individual path. Or of the main characteristics one notices is that some planets are surrounded by the atmosphere and others are without it. This will lead to significant consequences as regards temperature.

Living species are found to have existed on earth for a few billions of years (Calvin, 1969; Birdi, 2020). This is the most unique feature on the planet earth only (until present). The primary phenomena related to life on earth are attributed to the following physical-chemical characteristics:

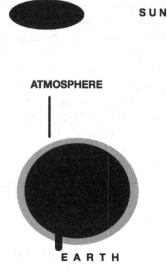

SUN

ATMOSPHERE

EARTH

FIGURE 1.7 Sun emits heat radiation on atmosphere – earth (solar radiation-heat is estimated as 340 Watt/m²: reflected heat radiation is 110 Watt/m²).

EXAMPLE

>PRIMARY phenomena:

>**SUN** – ATMOSPHERE – **EARTH** (LAND – FORESTS – PLANTS – FOOD – OCEANS – LAKES – RIVERS)

>SECONDARY PHENOMENA
>INTERACTIONS:
...**PHOTOSYNTHESIS** (BASED ON: SUN-LIGHT/CARBON DIOXIDE (CO_2)/WATER (H_2O)/PLANTS-FORESTS-FOOD/FISHERIES ($CO_{2,AQ}$ OCEANS)

$$[CO_{2aqeous}] == [CO_{2aq}$$

It is thus obvious that the process of *photosynthesis* (which is basically a process based on: < sunlight + carbon dioxide (gaseous) + water) is essential for the existence of life on earth (i.e. food – fisheries).

The process of *photosynthesis* occurs both on the surface of earth and to some extent in oceans/lakes/rivers (Birdi, 2020; IPCC, 2011, 2019; Rosenzweig et al., 2021; Manahan, 2022). The photosynthesis process has been occurring as follows:

I. During the pre-living species era.
II. During the living species era until now.

The kinetics of photosynthesis are variable and unpredictable (such as man-made activities).

In the present context, the primary aim is to consider the state of climate (as regards change of temperature), only with respect to any man-made industrial activity, around the planet earth. The heat received from sun (primary) on earth is obviously known to control the climate temperature conditions. In some recent studies (Birdi, 2020; Kemp et al., 2022), the worldwide equilibrium change of the temperature of earth was analyzed in more extensive detail (Appendix A).

Additionally, in some of these cases besides the latter, there is need to consider the effect of increasing degree of *pollution (air; drinking water; oceans/lakes/rivers).* It is well established that the term pollution refers, in general, to any man-made activity which interferes with the natural environment (   ).

At the preset, it is not quantitatively established, as regards the intensity of either of these processes (Appendix A). However, it is well established that increasing degrees of air pollution have resulted in among others higher earth temperatures (as observed in very congested and large cities (over 10 million people) worldwide.

As described earlier, the interactions between the sun and the planet earth are different and depending on the process by which this may be taking place. In the present, these stages are individually delineated for the sake of simplicity.

Furthermore, the stage in the previous case:

- SUN-----LAND-SCAPE---------FOOD (*CORN-WHEAT-RICE-ETC.*)
- SUN – OCEAN – FOOD (*FISH + SHELLS + ALGAE (FISHERIES)*)

(Basic process = photosynthesis) (plants – on land/in oceans-lakes-rivers)

These two different processes are life sustaining food essentials for living species.

It is thus useful to emphasize the need for food for the existence of all living species on the earth.

It is also important to notice that all fisheries' growth is dependent on the carbonaceous (i.e. available carbon dioxide (both in air (CO_{2gas}) and as dissolved in water (CO_{2aq}) (oceans/lakes/rivers)) (GCCSI, 2016; Birdi, 2020).

These different steps are very significant for the existence of living species on the earth. These both impact both strongly and essentially the need of carbon dioxide in air (at around: 400 ppm (0.04%) and oceans (in equilibrium with $CO_{2,gas}$ in the atmosphere)). At this stage, one needs to consider the mass balance of carbon dioxide in gas form. This is essential in order to consider the effect of any man-made carbon dioxide addition to the atmosphere.

In other words, the mass balance of carbon is under scrutiny in the present era (Birdi, 2020). The subject matter is simply delineated as follows (Appendix A):

EXAMPLE

MASS BALANCE OF CARBON DIOXIDE (CO_{2gas}) (gaseous: Earth Planet)

(Carbon dioxide exists only in the gas state at **RTP** (room-temperature-pressure)(App C)
>Different phases of Carbon:
>DIRECT $CO_{2,GAS}$
>Atmosphere: CO_{2gas}
>Oceans-lakes-rivers: (CO_{2aq} in dissolved state + carbonate cycle + shells)
>INDIRECT EXISTENCE OF CO_{2gas}
>FORESTS (PHOTOSYNTHESIS)(ADSORPTION/DESORPTION)
>OCEANS/LAKES/RIVERS (AS CO_{2aq})(ADSORPTION/DESORPTION)

In this stage of food-fisheries for living species (especially humanoids), the role of carbon (CO_{2gas}) becomes very fundamental interest. The carbon cycle can be described as follows.

EXAMPLE

(**ADSORPTION AND ABSORPTION IN OCEANS/LAKES/RIVERS** OF CO_{2gas})
        $SUN - OCEAN$ (CO_{2aq} + CaO = $CaCO_3$ (SHELLS/SKELETONS FISH/OTHERS)

Fisheries (food) is thus dependent on the availability of carbon dioxide in the atmosphere, which is in equilibrium with carbon dioxide (CO_{2aq}) dissolved in oceans/lakes/rivers (Appendix A).

EXAMPLE

<FISHERIES AND CEE
    <ALL FISH/SHELL/ALGAE IN OCEANS/LAKES/RIVERS
    PHOTOSYNTHESIS IN AQUEOUS MEDIUM
    <THIS NEEDS CO_{2aq} + WATER + OXYGEN + SUNSHINE
[Equilibrium Concentration of CO_2 in aqueous media = CO_{2aq}]

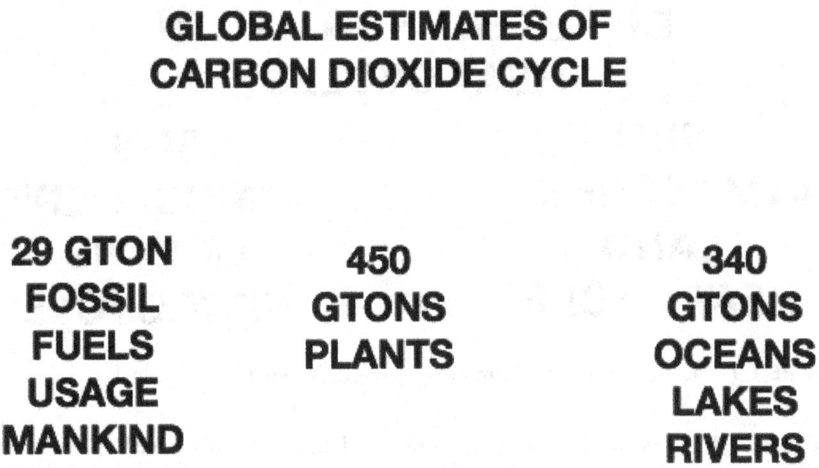

GLOBAL ESTIMATES OF CARBON DIOXIDE CYCLE

29 GTON FOSSIL FUELS USAGE MANKIND

450 GTONS PLANTS

340 GTONS OCEANS LAKES RIVERS

FIGURE 1.8 Carbon dioxide cycle around the atmosphere – earth (oceans/lakes/rivers and plants).

This needs more detailed and pertinent description about the role of carbon dioxide (gas) as found in air (atmosphere). There are found different components in air. However, only carbon dioxide, CO_{2gas}, interacts with water (e.g. Oceans/lakes/rivers), while none of the other components interact to the same degree (Appendix A). Furthermore, oceans cover about 70% of the surface of earth. Additionally, the depth of oceans can reach more than 5 km (5,000 m) in some areas. This shows that the solubility characteristics of CO_{2gas} in oceans can be of vast significance in the present context (Figure 1.8).

EXAMPLE:

>MASS BALANCE OF CARBON ($CO_{2,gas}$) AS FOUND IN ATMOSPHERE/ OCEANS/LAKES/RIVERS (EXCLUDING IN CAPTURED STATE IN PLANTS/ FOREST)

<ATMOSPHERE = 750 Gton $CO_{2,gas}$

<OCEANS/LAKES/RIVERS = >40,000 Gton CO_{2aq}

<(CO_{2gas}) << (CO_{2aq})

(Gton = 10^9 Tons)

This mass distribution analysis is found to be significant, especially, the chemical equilibrium (Calvin, 1969; Birdi, 2020).

In addition, there are other (besides man-made) carbon sink processes, such as photosynthesis, rain, and carbonates.

The quantity related to global (climate) heat is reported to be changing (nonlinear) and mentioned as climate change (Enns, 2010).

DIFFERENT SUN-EARTH
HEAT CYCLES

SUN	SUN
ATMOSPHERE	**ATMOSPHERE**
LAND	**OCEANS**
HEAT CYCLE	**HEAT CYCLE**

FIGURE 1.9   Heat cycles of sun – atmosphere – earth (on land or on oceans).

Current global climate *surface temperature* data are mainly based on some arbitrary fixed sites on the surface of earth (Birdi, 2020; Kemp et al., 2022; IPCC, 2011; Rosenzweig et al., 2021). They are used to average the change in the surface climate temperature variations of the earth.

In the system: sun – atmosphere – earth (with 30% land + 70% oceans), the heat radiation cycle gets more complex (Figure 1.9). The heat capacities (i.e. heat absorbed/reflected per unit area) are different for different substances (e.g. land/ocean). In other words, the heat input will vary differently on land or oceans.

It is important to mention that all these natural processes are known to be changing to nonlinear. Especially, this is considered to be related to the heat balance of the sun and earth (Birdi, 2020). Although it may not seem obvious, it must be mentioned that in the solar system, all planets are revolving around the sun (zz). This leads to nonlinear (Enns, 2011), interactions, especially as regards the varying magnitude of heat radiation input from the sun on earth.

EXAMPLE:

SOLAR PLANATERY SYSTEM:
   (Planets revolve around the Sun) (Paths (elliptical) are Variable)
   S U N (*VERY HOT* – EMITS HEAT RADIATION/DYNAMIC/*NONLINEAR EMISSIONS*)

P L A N E T S	(Distance from Sun)
MERCURY	(36 m.m/58 m.km)
VENUS	(70 m.m/108 m.km)
EARTH	(93 m.m/149 m.km)
MARS	(142 m.m/228 m.km)
JUPITER	(484 m.m/778 m.km)
SATURN	(886 m.m/1,400 m.km)
URANUS	(1,800 m.m/2,900 m.km)
NEPTUNE	(2,800 m.m/4,500 m.km)

(m.m = 10^6 miles; m.km = 10^6 km)

(ALL PLANETS RECEIVE HEAT RADIATION FROM SUN)

(Sun temperature: surface = 5,500°C; interior = >15 million °C)

S U N. >>>>>>>>>> ATMOSPHERE >>>>>>>>> E A R T H

INTERACTION:

(SUN – EARTH INTERACTION):

SUN (Heat Radiation) – Atmosphere – Earth (planet)(30% (land) + 70% (oceans + lakes + rivers)

SUN (HEAT-SOLAR RADIATION) – HEAT RECEIVED ON EARTH – HEAT REFLECTED FROM THE SURFACE OF EARTH

<CURRENTLY>: EARTH TEMPERATURE (SURFACE) IS REPORTED TO BE INCREASING DUE TO MAN-MADE INCREASES IN CARBON DIOXIDE (and other GHG gases) FROM FOSSIL FUEL USAGE FOR ENERGY PRODUCTION SINCE THE START OF INDUSTRIAL REVOLUTION (about 200 years ago)>

Furthermore, it is reported that most of these interactions are *nonlinear*.

One also finds that there are observations of different seasons on earth (summer) winter/rain-season). This thus indicates that climate and temperature are varying continuously (nonlinear) in time and space. This nonlinear behavior is becoming very urgent in current status around the world.

1. photosynthesis (pre-living species) (evolutionary equilibrium concentration of CO_{2gas} in atmosphere)
2. fire (wood burning) (additional production of carbon dioxide (CO_{2gas}))
3. fossil fuel combustion (pro-industry revolution) (production of CO_{2gas})
4. pollution (arising from man-made phenomena; from <1> and <2>)

It is thus obvious that any earth climate analyses need to be related to many related phenomena to any change in climate exclusivity (Sally, 1996; Kemp et al. 2022; Stute et al., 2001).

In general, one may expect that the average temperature on each planet varies, as dependent on various factors in relation to sun and other constraints. In the present case, subject matter concerns mainly about the state on planet earth.

The planet earth is known to revolve around the sun (solar system). Life on earth has existed for many geological (about 4.6 billions) years. The reason and basis for the existence of life (solely) on earth are not absolutely clear. However, there are various factors which indicate the reason for the latter phenomenon. The solar system consists of various planets which are known to revolve around the sun. It is known that life only exists on planet earth (life as one knows on earth) (due to the photosynthesis phenomena (which requires: carbon dioxide + water + sunshine). Planets in this solar system are found to exhibit properties which are complex systems. Some planets are made of only gases, while others are solids (such as earth (only crust); moon; Mars) (Appendix A).

Earth is part of the solar system, with sun being the main star. This system will be discussed only in simple terms, since some properties are essential in the present context.

Due to earth's gravitational forces, some gas molecules are attracted (which is termed as the atmosphere).

The atmosphere surrounding the earth is known to play a very significant role as regards the status of various conditions on earth:

EXAMPLE

(BASED ON SPECIFIC PHYSICAL PARAMETERS: TEMPERATURE/ ATMOSPHERE/PRESSURE)
>WATER
>LOW CARBON DIOXIDE (CO_{2gas}) CONCENTRATION (400 ppm:0.04%) AS GAS IN THE ATMOSPHERE
>SOLAR PHOTOSYNTHESIS (COMBINATION WITH CO2/WATER (H2O)

These parameters have reached an equilibrium (pseudo) over the evolutionary period since the creation of earth (as a hot lava-like) planet to present state with crust on the surface (Appendix A).

Based on unknown evolutionary conditions (for obvious reasons), after the earth partially cooled at the surface, the atmosphere was established where gas molecules were held surrounding the earth (as atmosphere) over a few billions of years. The composition of the atmosphere is found to have stabilized, such as to provide an element for the evolution of life (living species) on earth over these eras.

In order to describe the planet earth, in very simple terms:

- it cooled down some billion years ago; to form a crust on its outside, while the interior is known to remain hot and liquid-like (lava) (Birdi, 2020).
- Simple living species appeared around this time.

EXAMPLE:

<EARTH (temperature): OUTSIDE AND INTERIOR:

>COOLING OF EARTH IN THE FORMATION OF CRUST (Started 4.6 b.years ago)
>INTERIOR OF EARTH CURRENTLY VERY HOT AND LIQUID-LIKE (evidence from earthquake lava eruptions)
>EARTHQUAKES ERUPTIONS INDUCES COOLING

This simple observation suggests that:

- earth crust had started to form in early stages of the creation of the solar system (due to cooling) (and observations indirectly indicate that this process is ongoing: earthquakes).
- this process is very complicated and no clear/plausible definition on this temperature of climate change has been found in the literature.

Earthquake eruptions on earth are known to give rise to various inputs to climate. These can briefly be considered here for the sake of simple parameters (Appendix A). The qualitative factors which may be mentioned are as follows:

- hot lava (from the interior of earth) erupts toward the colder crust (heat is added from the hot interior to the surface (cool) of earth)
- it adds various gases (pollutants) to the atmosphere.
- it adds considerable pollution due to dust particles which are observed worldwide.
- earthquakes are normally erratic and unpredictable.

The earth is surrounded by atmosphere (air) and water is found in great oceans (covering almost 75% of the surface crust of earth). It is also suggested that the sun + earth interactions are controlled by the presence of the atmosphere. The latter is one of the main subject matters in the present context.

The molecular density of the atmosphere varies with height from the earth.

EXAMPLE

<EARTH-ATMOSPHERE (AIR) LAYERS:

ATMOSPHERE (estimated (average (Temperature/Height))
STRATOSPHERE--------
EARTH-----(LAND-OCEANS-LAKES-RIVERS-FORESTS-PLANTS-
    FOOD)

Furthermore, the earth – atmosphere – sun system:
SUN-----ATMOSPHERE-----EARTH

---

comprises a constellation which man has admired and studied for many centuries.

However, the temperature in the solar system has had a very significant evolutionary phenomena (Appendix: A).

These natural aspects have been studied by man for thousands of decades. The degree of information has increased with innovations.

During the past, different religions and civilizations have worshipped certain specific natural elements for thousands of years. Some of these *elements* were/are (some religion still worship) as follows:

- Air (atmosphere; wind; rain)
- Water (rivers; lakes; oceans)
- Sun (light: heat: radiation)
- Fire (flame; combustion)

Atmosphere-carbon dioxide-rain: This system plays an important role in the current context. CO_{2gas} is known to be soluble in water. The interface of atmosphere (CO_{2gas}) – water (rain drops) behaves as natural carbon sink. The ratio of surface area of rain drops to the atmospheric interface is very large.

Thus, rain drops are known to absorb CO_{2gas} and fall on earth as CO_{2aq}. This phenomenon is known to affect the infrastructure of earth in different ways (e.g. monuments; calcification;).

In this context the component water is obviously related to water as found in the liquid state and vapor (in atmosphere: clouds-rain drops).

The basic life sustainability on earth is primarily dependent on the availability of food (corn, wheat, rice, fruits, fisheries, etc.). In a different context, the concentration of carbon dioxide (CO_2: also termed as carbon) in air becomes very important in more than one aspect. This is related to the existence of mankind/mammals. The latter has various implications, as described in this study.

EXAMPLE:

SUN – (160 million km) – *ATMOSPHERE* – (20 km) – EARTH
>SUN=EMITS HEAT RADIATION
<ATMOSPHERE OF EARTH = HEAT FROM SUN IS REFLECTED/ABSORBED/TRANSMITTED
AT THE *INTERFACE*
>EARTH = CONSISTS OF OCEANS/LAND/*FORESTS/PLANTS/FOOD*
<EARTH = ABSORBS/REFLECTS
>EARTH GLOBAL TEMPERATURE = NONLINEAR SINCE COOLING-ERA AND VARIABLE
EARTH >>GEOTHERMAL HEAT FROM INTERIOR CORE

In short:

1. sun radiation travels about 160 million km, and interacts with the interface of atmosphere.
2. heat radiation from <1> is transmitted under dynamic conditions and interacts the surface of earth (which is spherical).

1.2.1 ATMOSPHERE AND CLIMATE CHANGE

<This study relates to the carbon (i.e. carbon dioxide: CO_{2gas}) (term carbon is invariably used instead of carbon dioxide) capture and **surface chemistry** properties of **carbon dioxide** (CO_2), especially its connections to the change in temperature (average surface temperature) of earth. However, it is useful to describe some general remarks about the atmosphere (air) and its characteristics (especially with respect to surface chemistry).

The composition (and temperature and pressure) and density (molecule per unit volume) of air vary with height over earth. In other words, when one considers the composition of air, the height has to be considered. The gradient of composition of air (i.e. variation of composition with height), might be dependent on time-scale (and space). The density of air decreases as height increases. This arises from the fact that the molecular attraction forces decrease. The air density can be determined from the knowledge of height and temperature and pressure (Appendix B).

Similarly, water (e.g. oceans; rivers; lakes; drinking water) contains a variety of different salts (sodium (Na), calcium-ion (Ca^{++}), magnesium (Mg^{++}), chlorides (Cl^-), sulfates (SO_4^-), etc.), which are known to interact with some components of air (for example: carbon dioxide (CO_{2gas}); oxygen (O_2)). Mankind has been aware of the essential role played by these elements for the existence of life on earth. Even though millions of miles away, the sun provides heat radiation to the planet earth. Besides heat, sun also emits other kinds of energy to earth, e.g., radiation and ultraviolet light. The atmosphere surrounding the earth has four distinct layers (Sally, 1996; Birdi, 2020):

EXAMPLE:

<ATMOSPHERE STRUCTURE AND ITS RELATION TO EARTH:
>Relation to height from the surface of earth:

<>0 to 10 miles (0 to 16km: temperature range ca. 20°C to −50°C: **TROPOSPHERE**: This is the region of human activities.

<>10 to 30 miles (16 to 50km: temperature range ca. −50°C to 0°C: **STRATOSPHERE**: The ozone gas (O_3) layer is found in this region, which absorbs the ultraviolet (UV) radiation from the sun. The temperature is higher at a higher altitude, due to the absorbance of ultraviolet radiation from the Sun.

<>30 to 50 miles (50 to 80km: temperature range from ca. 0°C to 90°C): **MESOSPHERE**: In the mesosphere, temperature decreases as altitude increases.

<>50 to 75 miles (80 to 200km: temperature range from ca. >90°C): **THERMOSPHERE**: In this region temperature increases with altitude, due to absorption of highly energetic solar radiation. It becomes even more evident that the system is nonlinear as regards heat flux.

The planet earth is surrounded by the gaseous atmosphere. Therefore, the current analysis based on fixed sites for temperature measurement is insufficient, since there is a need for gradient analyses. Again, one must consider the infinite parameters in natural phenomena. The latter thus restricts the analyses without simplifications and data simplifications used in any model (Kemp et al., 2022; models).

Molecules in any gas (air) phase are attracted to the earth. As the force of gravity weakens with increasing height, molecules in air become scarce as the atmospheric pressure drops. This leads to vacuum at very high distances. The distance between gas molecules (mean-free path) (Knudsen, 1917; Birdi, 2020) increases to very large values at heights around 20 km (high vacuum effect).

Furthermore, there are two main significance distinct kinds of earth surfaces; e.g. land (20%) or oceans (75%) (i.e. water). This means that the interfaces of air – land and air – oceans are highly different in size. In addition, the oceans can be over 5 km in depth invariably. *It is recognized that most of the heat absorbed the earth, from sun, is in oceans.* This means that most of the heat from sun will be absorbed in the oceans (Kemp et al., 2022).

EXAMPLE:

INTERACTIONS BETWEEN AIR (ESPECIALLY CO_{2fas}) WITH DIFFERENT SURFACES ON EARTH:

AIR – OCEANS ($CO_{2,air}$ – OCEANS)
AIR – LAND (DESERT OR ARID)
AIR – FORESTS-PLANTS
AIR – POLAR ICE-CAPS

This clearly shows the different paths of sun – atmosphere system and the interactions relating to temperature on the earth (Birdi, 2020; Kemp et al., 2022).

Further, man has used **fire** (with respect to: food; combustion processes; energy (electricity; mechanical); transportation) in many different applications and technologies (both directly and indirectly) for many hundreds of years. It is also recognized that all these elements are essential for the existence of life on earth.

1.2.2 GREENHOUSE GAS (GHG) AND CLIMATE CHANGE (SURFACE TEMPERATURE)

The basic criterion subject of this manuscript arises from the fact that the quantity of heat reaching the earth (and its atmosphere of gases), from the sun is through the long wavelength of the light (infrared) (Calvin, 1969; Birdi, 2020) (Appendix A). The term greenhouse gas (GHG) is related to the molecular absorption of a substance in the infrared region. In other words, if the light is absorbed in this region of light, then heat is absorbed. There are various molecules in air which are known to absorb sunlight in this region.

EXAMPLE:

<Various Greenhouse Gases As Found In Atmosphere (Earth-atmosphere):

METHANE GAS (CH$_4$) – CARBON DIOXIDE GAS (CO$_{2gas}$) – NITROUS OXIDE GAS (N$_2$O) – FLOURINATED GASES (FLOUROCARBONS)

Water Clouds Vapors (H$_2$O)---
[Air-pollution is not GHG; but it contributes to global warming.]
It is useful to give a brief description about each GHG gas. In the atmosphere, some of the gases, which appeared under the evolution, are as follows.
FOR EXAMPLE:

- <methane: Natural gas reservoirs are currently found in the earth. This gas can diffuse through cracks to the surface of earth. However, this has very low concentration.
- <Carbon dioxide (CO$_{2gas}$): Photosynthesis has been observed to have existed pre-living species. Photosynthesis phenomena (petrified plants) indicates the need of CO$_{2gas}$. Thus CO$_{2gas}$ has been in the atmosphere for billions of years (Calvin, 1969). Currently, it is 420 ppm and in equilibrium with CO$_{2aq}$ in oceans. In other words:
  - *photosynthesis has been taking place for billions of years.*
  - *GHG phenomena have thus also been present for billions of years.*
  - both photosynthesis and GHG have thus attained a pseudo equilibrium at present

EXAMPLE

<GHG EFFECT AND EVOLUTION
  <PRE-WATER::NO ATMOSPHERE::NO GHG
  <CLOUDS::PRE-INDUSTRIAL::ATMOSPHERE::WITH GHG
  <TODAY::WITH INDUSTRIAL::WITH GHG=CLOUDS-CO$_2$

The climate change of the surface of earth is suggested to be dependent on different physical factors (as reported from evolutionary processes) (Kemp et al., 2022; Lomborg, 2022) (Appendix A).

This is known from evolutionary observations (Calvin, 1969). Most significant observation is based on the fact that all normal humans have about $37°C \pm 1°C$, temperature, worldwide, regardless of time/place/food/climate (Birdi, 2020). This is mainly based on the body temperature reaching *equilibrium* within the various biological processes (Appendix A). The quantity body temperature is found to be independent of surrounding temperature (Birdi, 2020).

This example is not directly related to the current subject, but a similar-example is pertinent. All living bodies are known to stabilize at some norm temperature.

The degree of absorption in the infra-red, by any molecule, is dependent on its molecular structure. Thus different molecules exhibit varying degrees of absorption.

EXAMPLE

<COMPARATIVE DEGREE OF GHG OR ABSORPTION…

Gas Molecule	GHG (relative) EFFECT
METHANE (CH_4)	25x
CARBON DIOXIDE (CO_{2gas})	1x (arbitrary)
Nitrous oxides (*NO_x)	295x
Fluorocarbons	8000x
WATER VAPOR (CLOUDS)	(reflection)

----(CO_{2aq} exhibits no GHG properties.)

These data show that the degree of GHG is different for different gases (pollutant). The concentrations of different GHGs are also varying. Furthermore, these GHG molecules have reached an evolutionary equilibrium at the present stage.

1.2.3 SOURCES OF HEAT EMISSIONS AND SINKS IN EARTH SURROUNDING

It is also necessary to briefly consider the earth's heat balance with its surroundings (energy sources or energy sinks) which are needed for consideration (Appendix B):
SOURCES OF HEAT:

- SUN…(major)
- Mankind combustion technologies:

  1. pre-industrial revolution (mainly wood combustion)
  2. after industrial revolution (200 years ago) (crude oil (after refinery) + natural gas + coal)

- Diverse sources of energy (plants, forests, storms)
- Earth core (temperature is very high) (dissipation of heat) (earthquakes)

SOURCES OF HEAT SINK:
* Evaporation of water (from oceans (northern/southern hemispheres are two different states, lakes, rivers, etc.) to make clouds/trees (which reflect sun heat)

1.2.4 DYNAMICS OF TEMPERATURE OF EARTH

It must be mentioned that there is no parameter (e.g. temperature, pressure, and wind velocity) which is static at any point of observation on the surface of earth (i.e. with regard to time and place). All parameters, as found in environment, are varying non-linearly, with time and space on the earth. In other words, all evolutionary processes are and have been nonlinear (Enns, 2010).

This observation thus indicates that throughout the evolutionary process on earth, different dependencies of matter have been nonlinear and geological long range predictions are not as straightforward as one would like to expect.

For example, on the average, on a typical day, the temperatures (near the surface of earth) at any place on earth are higher during the daytime than in the night:

Temperatures at different parts on earth (day/night temperature) (on April 3, 2023)

CITY.................................DAY/NIGHT TEMPERATURE (°C)
....................(DATES).....MAY-2018.........JANUARY-2019

CITY	DAY	NIGHT
Copenhagen	6	−1
London	12	5
Frankfurt	7	−1
Livigno	−2	−9
Moscow	13	5
Calgary	3	−2
Delhi	31	17
Nice	18	10
Buenos Aires	21	17
Malaga	23	11
New York	11	4
Manila	29	20
Berlin	6	0
Rome	20	9

In addition, the temperature of oceans also changes with day/night/summer/winter parameters. The day/night temperature data merely show the heat (dynamics) sun adds to the earth at different places (and time) on the earth. It is accepted that in sun – oceans interaction, the heat absorbed by the oceans is a very large quantity. It is known that the density – temperature variation of water is non-ideal. The density of water shows a maximum at around 4°C (Birdi, 2020). This indicates that if cooled to 4°C, it will sink to the bottom of ocean. This creates a natural water-ocean circulation phenomenon.

These observations indicate the temperature fluctuations existing at every instance around the clock. This indicates the huge magnitude of sun energy (which fluctuates day/night) reaching the earth. It is obvious from these data that any analyses of average temperature of earth would be rather complex (Kemp et al., 2022).

Especially, considering the fact that man has no control of what processes are prevalent on the sun and the changes which may have a profound effect on the climate on earth.

In the same context, the heat reaching the earth from sun is also different with time and space. The rotation of earth also contributes to the dynamics of temperature at any given time or space. Earth axis is tilted toward sun. The angle of tilt varies with time and space.

Further, the dynamics of temperature is known to create weather turbulences. The latter phenomenon indicates that weather changes (as regards temperature, winds, waves, storms, etc.) both due to differences in heat at night/day, but also other related phenomena over weeks/months/years.

1.2.5 ATMOSPHERE OF THE EARTH AND THE SUN

Atmosphere surrounding the earth is a very important element as regards the life on the earth. The composition of atmosphere (air) and its temperature (and pressure) is found to vary with height from the surface of earth. The sun energy gives rise to heat on the earth (see day/night temperature). The atmosphere (and clouds) controls the amount of heat reaching the earth in some complicated paths (e.g. absorption and reflection).

The planet earth (surrounded by the gaseous atmosphere) is known to receive heat from sun (surface temperature 5,000°C).

Sun is ca. 92 millions of miles (160 million km) away from the earth. Sun energy (as heat) is found to be thus enormous, as compared to man-made processes (Appendix A). The degree of sun shine reflected by the atmosphere is also a very important element. The degree of reflection by the atmosphere is also dependent on the height (from the surface of earth). The surface temperature of the sun is about 5000°C.

This study relates to the surface chemistry (a special theme in general chemistry) (Chapter 3; Appendix A) and its relation to control/storage/recycling of CO_2 in atmosphere (current concentration $= 0.04\% = 400$ ppm). The evolutionary chemical evolution (CEE) has created the current state of the environment on earth.

Over the past few decades (especially after the industrial revolution) the concentration of CO_{2gas} is found to be increasing (Figure 1.2). (IPCC, 2019; Tziperman, 2022): However, beginning from:

- pre-living species age,
- photosynthesis and GHG age

This geological system: CO_{2gas}/photosynthesis/GHG has reached pseudo-equilibrium (Figure 1.10)..

TYPICAL CO$_{2gas}$ (PPM) DATA: (YEAR/PPM):

- 1960/310 ppm
- 1975/330 ppm
- 1980/340 ppm

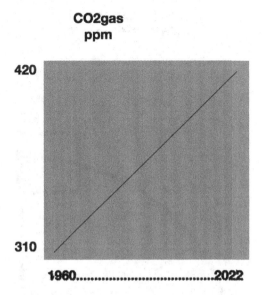

FIGURE 1.10 Change of CO_2 concentration in air (near the surface of earth).

- 1995/360 ppm
- 2010/390 ppm
- 2017/410 ppm

Furthermore, the usage of fossil fuels has varied from the western countries toward the eastern (e.g. China/India) developing countries over the few past decades.

This observation is attributed to the increasing use of fossil fuels (primarily proportional to increasing world population and the technology) (Appendix A). Currently, global consumption of different fossil fuels is ca. 33% each (equivalents to 100 million barrels/day oil) (Figure 1.11):

OIL + COAL + NATURAL GAS.

EXAMPLE

>FOSSIL FUELS TYPICAL USAGE (WORLDWIDE) ESTIMATES:

>*1< OIL (CRUDE) (30%)+NATURAL GAS (30%) + COAL (30%)*
= *total 300 million barrels oil equivalent/per day*
= *total CO_{2gas} man-made = 300 10^6 159 liter oil 1 kg CO_2 = 48 10^9 kg CO_2 per day*

(World population 2023 -<> 8 billion = 8 10^9)
This is already verified in some regions. For instance, the population in China is already (2023) stagnant (at about 1.4 billion).

CLIMATE CHANGE MODELS
(FROM 1960 to 2050)

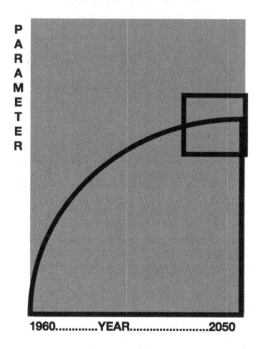

FIGURE 1.11   Climate change models and predictions of various parameters (e.g. fossil fuel used by mankind; energy used per capita; growth of worldwide population; availability of fossil fuels; economic worldwide growth) (Appendix A). The square area indicates the degree of estimation.

At the energy/power factories, where fossil fuels are used (after treatment at refinery), the exhaust flue gases contains ca. 10 % CO_{2gas} as compared to 0.04% (400 ppm) in air. In order to mitigate the increasing CO_{2gas} the capture of the latter (by application of CCRS) has been used. In-fact, it has been recognized that application of CCRS is the only viable technology to mitigate the carbon capture process and the control of climate change (Chapter 4). This book describes the technology and surface chemistry of the carbon capture recycling and storage (CCRS). The main approach is to present a review of the CCS technology as found in the current literature.

1.2.6   GREENHOUSE GASES (GHGS) IN THE (ATMOSPHERE) AIR

Climate change (temperature increase over the past few decades) on earth (surface) is reported related to the greenhouse gas (GHG) emission (IPCC, 2011, Rackley, 2010; Kemp et al, 2022; Lomborg, 2022; Birdi, 2020).

Methane (as natural gas reservoirs) and CO_{2gas} are found in nature as free gases in atmosphere. Carbon dioxide, CO_{2gas}, is also produced by all fossil fuel combustion (and fire) processes by mankind. However, NO_x and F_xH_y are man-made pollutants and present in almost all flue gases.

It is important to mention the most significant role played by sun, as regards the heat radiation (temperature) balance of earth. The mass of sun is almost 95% of the total solar system. Sun is mainly composed of hydrogen gas (92%), plus about 8% of helium and other elements (Manuel, 2009).

The magnitude of heat radiation reaching the earth from sun is very significant (estimated: 350 watt/m^2). It also becomes obvious that this quantity is varying due to the continuous movements of the planets earth/sun.

The heat radiation reaching the earth from sun, via atmosphere (via: clear day/cloudy/rainy/storm), is very dynamic. The daily cycle (i.e. sunrise/sunset) imparts dynamic heat transfer to the earth. Further, sun surface is found to show flare activity, which is variable over time. The latter is also found to be variable. Further, the clouds in the earth's atmosphere reflect the heat from sun. Since there are different kinds of clouds, the degree of reflection of sun rays is dependent on this latter parameter. All in all, one thus finds that the heat transfer from sun to earth is highly variable from day to day (as well as in time and space). In addition, one must mention that the difference in temperature on earth from day night can be as high as 20°C. Further, the solar energy reaching the earth consists of different wavelengths (as seen from the diffraction through a prism):

The range of visible light from sun is from 380 to 750 nm (Birdi, 2020):

<VIOLET...............IINDIGO...........BLUE............GREEN..................YELLOW ....ORANGE ....RED>

<380.. ....750> (WAVELENGTH: nm: 10^{-9}m)

EXAMPLE:::

SCHEMATICS OF NON-LINEAR PARAMETERS AND TEMPERATURE INVOLVING EARTH

SOLAR (*VARIABILITY*) – ATMOSPHERE (*NON-LINEAR COMPOSITION*) – (*GHG-NON-LINEARITY*) – INTERFACIAL SURFACES (ATMOSPHERE-LAND/ATMOSPHERE-OCEANS/LAKES/RIVERS–ATMOSPHERE-PLANTS(FORESTS)

Nonlinearity in these systems will be expected to give rise to unexpected variations under the evolutionary process (Enns, 2010). However, the significant observation that living species on earth have continuously survived over the past millions of years only supports that the nonlinearity has been of no negative consequences.

It is useful to explain the current description of the term *climate change* of the surface of earth.

The term **climate change**, i.e. average global temperature measurement on the surface of earth mechanism would thus be expected to be rather a complex phenomenon (Kemp et al., 2022). For many decades, it was suggested that the heat (infrared) from sun is absorbed by GHGs, and this is expected to induce an increase in temperature of the earth. GHG gases (such as: CO_2; CH_4; NO_2; Perfluorocarbons;) are molecules in air which absorb infrared light. This leads to heat absorption from the reflected sun energy falling on the earth (Li et al., 2011). It is estimated that the average power density of solar radiation at the outside of the atmosphere of earth is approximately 1,366 W/m², which is called the Solar constant. Further, it is found from these data and the surface area of earth that annual solar energy reaching the earth is ca. 5,460,000 EJ/year. It is important to compare this with annual energy consumption by man, which is about 500 EJ. This means that only 0.01% of solar energy is what mankind consumes/emits. The concurrent increase in pollution with any mankind activity is also known to contribute to increase in temperature (Birdi, 2020; IPCC, 2022).

1.3   COMPOSITION OF GASES IN ATMOSPHERE (AIR)

Sun emits heat radiation (about 350 watt/m²) and it interacts at the *interface with* the atmosphere surrounding the earth. The composition of gases in atmosphere has reached an equilibrium, atmosphere extends about 10 km above earth, while decreasing in molecular density. Above 10 km height, molecular density is too rare for living species (without extra oxygen). In other words, the GHG effect is also dependent on the molecular density. Most of the GHG effect will be expected around 1–2 km from the surface of earth.

It is reported that about 30% of solar radiation is reflected from atmosphere interface, from the sun – atmosphere interface.

The atmosphere surrounding the earth consists of different gas molecules. The composition of gases has reached a pseudo equilibrium during this chemical evolution (Calvin, 1969).

These gas molecules are attracted to earth by van der Waals forces (Appendix B). Since vdw forces decrease with distance from the surface of earth, the density of air drops to almost zero at about a height of 50 km (IPCC, 2022). In other words, the effect of any parameter will decrease with the height.

This shows that the GHG effect decreases nonlinearly. In addition, the difference in molecular size of different gases is also important. Due to physical characteristics, larger gas molecules tend to increase in concentration as the height increases ( ). This physical property of the heavier molecule, such as carbon dioxide, CO_{2gas}, and molecular weight of 44 gm/mole, will sink.

EXAMPLE

<Density of molecules in atmosphere as attracted to earth versus height (schematic).

    Surface.......................
    5 km. . . . .
    10 km. . . . .
    20 km.   .   .   .   .

Although it is not completely clear in the literature, evolutionary phenomena leading to various aspects in natural surroundings in earth have evolved the existence of living species on earth (Calvin, 1969).

All living species on earth (on land and oceans (fisheries)) are dependent on the composition of gases in atmosphere (Appendix A). Especially, comparison of different atmosphere of solar planets shows that their compositions are significantly different. The composition of earth atmosphere indicates the presence of comparatively low concentration of carbon dioxide. In contrast, atmosphere composition of other solar planets does not indicate this analysis. Furthermore, living species are found on earth to sustain by utilizing the low content of carbon dioxide in air: for existence. In other words, evolutionary living species have developed this phenomenon.

Therefore, it is of interest to mention some chemical aspects of the composition of air.

The average composition of air (or the atmosphere) as known today is a mixture of different gases (near the surface of earth):
ATMOSPHERE EQUILIBRIUM COMPOSITION (near surface of earth (at STP))
(Evolutionary equilibrium):
(**PURE GASES**):
Nitrogen (N_2: 28 gm/mol) .... (78%);
Oxygen (O_2: 32 gm/mol) ....(21%);
Argon (Ar: 40 gm/mol) ....(0.9 %);
Carbon dioxide (CO_2: 44 gm/mol) ....(0.04%); (400 ppm)(measured near surface of earth);
Water (H_2O: 18 gm/mol) (vapor)............. (traces);
Other gases (hydrogen, H_2)..........................(traces)
(presence/absence of particles is excluded: see Appendix pollution)

In addition, one finds varying types of pollution substances around the earth:

• dust particle pollution (especially around large cities)
• uncontrolled dumping of gaseous material into environment

The concentration of water (as vapor/clouds) is dependent on the temperature of oceans/lakes/rivers. The phenomenon of wind, storms, etc. is found to have a major effect on this. Many of these major phenomena are rather unpredictable to mankind, and thus, any past/future predictions are sometimes incorrect. These phenomena vary from time and place.

In the present case, the subject matter is the consideration of CO_{2gas} for mankind, especially in the context of effect on climate-temperature change on earth.

Various studies have indicated that the *equilibrium concentration of carbon dioxide* (CO_{2gas}) in the atmosphere is noted to be increasing since the advent of industrial revolution (and subsequent increases in production of $CO_{2,gas}$ from fossil fuel combustion) (Salty, 2012; Birdi, 2020; Kemp et al., 2022; Lomborg, 2022).

It is known that the carbon dioxide, CO_2, molecule is gas at RTP (room temperature and pressure). The carbon atom is covalently double bonded to two oxygen atoms. This is found to have a linear structure: $O=C=O$.

It is a transparent gas, but is known to absorb in the infrared radiation spectra. In other words, the heat reflected from the earth, which is infrared, gets trapped in the atmosphere. This has been termed as greenhouse gas (GHG). It is non-toxic at concentrations below 90.000 ppm. Its current concentration in air is 420 ppm. In comparison, in air, oxygen (28%) is no-toxic, while nitrogen (72%) is toxic if present without oxygen.

Most important aspect of this is that carbon dioxide is the only gas in atmosphere which is soluble and interacts with water in oceans at a much higher degree than the other so-called inert gases (oxygen (O_2); nitrogen (N_2)).

EXAMPLE

>Mass-balance (estimates) and cycling of carbon dioxide (gaseous) (CO_{2gas}):

$CO_{2,air}$ == 750 GtC
$CO_{2,oceans\,+}$ = 39,000 GtC

RATIO: 1:50
(This is pseudo-equilibrium state: GtC = G ton carbon (C)))

Obviously, one finds that this is a rather complex phenomenon (Kemp et al., 2022). Besides this, it will be attempted to explain the subject matter in some regulation manner. There exist various explanations in the current literature about this matter. These will be evaluated in very direct context. Furthermore, the subject matter will be presented to the reader.

The global average temperature (measured on the surface) of the earth is primarily regulated by the heat received from the sun (Appendix A). Furthermore, the energy (heat; radiation) reaching the earth from the sun has to pass (interact) through the atmosphere surrounding the earth. This means that the sun energy (radiation) has to interact with (Birdi, 2020; Kemp et al., 2022):

- gas molecules (and dust particles) in the air (atmosphere)
- reflection by air/clouds/
- reflection from earth, trees, and ocean surface

It becomes immediately obvious that the nonlinear heat balance is complex (Birdi, 2020; Kemp et al., 2022; Vallis, 2011). The nonlinear dependence of various factors will obviously make any model analysis very dubious and insecure (Kawamiya et al., 2020).

Especially, the nonlinear behavior is dependent on time and space: the latter has been discussed in the literature (Lomborg, 2022; Kemp et al., 2022).

Furthermore, since this has major correlation with the evolutionary aspects on earth, one needs to mention some of these parameters in this case. At each interface, e.g.:

- sun radiation – atmosphere interface
- atmosphere heat – land (earth) interface
- atmosphere heat – oceans (earth) interface

(Earth Surface: Oceans=70%/Land=30%)
    Each *arrow* in Figure 1.12 indicates the heat radiation path:

- transmission
- absorption
- emission

These phenomena are found to be nonlinear. The earth receives heat radiation and light from sun in varying amounts throughout time and space (summer/winter/day/night). The varying heat radiation input from the sun (the surface temperature of sun

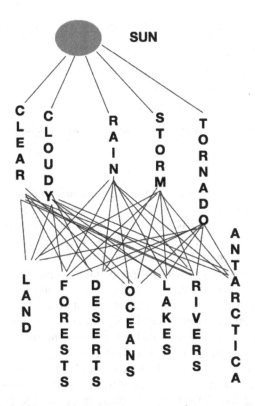

FIGURE 1.12 Various heat radiation transport routes from sun to atmosphere (e.g. clear/cloudy/rain/storm/tornado) to the surface of earth (where: 30% is land; 70% is oceans/lakes/rivers) (schematic).

is 5,500°C) is transmitted by the infrared wave length of the radiation. The infrared energy is absorbed by some gases as found in the atmosphere (e.g. carbon dioxide and methane). These gases are called *greenhouse gases* (GHGs). The absorbed infrared energy is reflected in all directions, which gives rise to increase in temperature of the planet earth. In the present text, there is only interest in **carbon dioxide** and its interactions with the temperature balance of the earth. Furthermore, it is known that the planet earth was originally formed as in liquid state at very high temperature. One finds evidence directly from the fact that earth is currently fluid lava in the middle of earth. This is occasionally seen during the earthquake eruption where lava moves from the inside to outer surfaces of earth. It is known that the largest effect of earthquakes has been on the air pollution on the environment.

Mankind has contributed in different ways to the changes in environment of earth. In the present case: the effect of fossil fuel combustion (thus production of increasing carbon dioxide (CO_2)) and its impact on climate change (arising from increasing pollution) (IPCC, 2011) (Appendix B).

EXAMPLE

>CLIMATE AND FOSSIL FUEL COMBUSTION AND POLLUTION:
  >ONLY POLLUTION EFFECT:
    >AIR-POLLUTION
    >DRINKING-WATER POLLUTION
    >OCEANS/RIVERS/LAKES POLLUTION
    >AND OTHER DIVERSE POLLUTION PHENOMENA

---

becomes obvious that any analysis regarding the concentration of carbon dioxide (CO_2) becomes ambiguous when the phase is air (atmosphere) or in oceans/lakes/rivers. Currently, though, one has proceeded by following the change in connection of CO_2 at a fixed place on earth. This only provides the concentration of CO_{2gas} in air, while CO_{2aq} concentration in oceans is seldom mentioned.

Currently, living species (all kinds) are only found on planet earth. Furthermore, mankind has developed in different ways to make various changes in the natural surroundings. The effect of climate change on the future existence of life (all living species) becomes significant when considering the man-made modification on earth.

In the current literature, one finds that the average concentration of carbon dioxide (CO_{2gas}) in air is reported to be increasing over the past century (ca. 2 ppm/year) (IPCC, 1996, 2011; Dubey et al., 2002; Dennis et al., 2014; Hinkov et al., 2016; Birdi, 2020). This has been attributed to the fossil fuel burning (combustion) (man-made CO_2) (Leung et al., 2014; McDonald et al., 2015). It is estimated to add about 2 ppm CO_{2gas} every year, over the past 200 years.

Furthermore, the increase in population and demands for increased energy leads to increase in man-made carbon dioxide (related to increasing fossil fuel combustion).

1.4   FOSSIL FUEL COMBUSTION PHENOMENA: FOSSIL FUELS + OXYGEN (FROM AIR) = FIRE (HEAT, ELECTRICITY, MECHANICAL ENERGY, ETC.) ... (CO$_{2GAS}$ PRODUCED)

However, the term *oil* (crude) combustion process needs more analysis. This is always shipped to various oil refineries for further separation into specific components. World population growth induces need from increasing magnitude of energy (IPCC, 2011).

EXAMPLE

<WORLD POPULATION INCREASE

  <1960 -> 3 Billion

   <2020 -> 8 Billion

    >2030 (estimated) -> 11 Billion

(billion = 10^9)
EXAMPLE: PROCESS OF COMBUSTION IN AIR.
FOSSIL FUEL = WOOD/COAL/OIL/GAS
AIR = OXYGEN (80%) + 19% NITROGEN + REST

The combustion process is of interest in the case of climate change and the role of CO_2 (Birdi, 2020; IPCC, 2011). These can be explained in the following. Fossil fuels, such as natural gas (CH_4), are known to react with oxygen (from air) as:
Natural Gas (CH_4 (>95%):

$$CH_4 + O_2 = CO_{2gas} + H_2O$$

OIL (after oil refinery) (paraffins: $C_n H_{2n+1}$)

$$C_n H_{2n+1} \text{ (crude oil)} + O_2 = CO_2 + H_2O + \text{other (pollutants: } NO_x; CO_2)$$

Wood:

$$\text{Wood} + O_2 = CO_2 + \text{other}$$

Coal:

$$\text{Coal} + O_2 = CO_2 + H_2O + \text{other substances}$$

(term other = minerals + pollutants)
[Crude oil is always refined into various fractions; after that it is used in different applications].

1.5 CARBON COMBUSTION::CRUDE OIL RESERVES AND REFINERY::PLASTICS AND RECYCLING

CRUDE OIL RESERVES and OIL REFINERY: It is useful at this stage to describe the term fossil fuel oil. The recovery and usage of oil from reservoirs is a complex process. In general, all oil recovered is shipped to different *refineries* around the world. Crude oil as one recovers from reservoirs is seldom used directly in any application. Instead, crude oil is shipped (currently ca. 100 million barrels per day) to different refineries worldwide. It is thus important to notice that refineries play a very vital role in the usage of crude oil in everyday life.

Each refinery has a specific operation, as determined by the type of crude oil and the area of refinery. In general, at a refinery the crude oil is separated into different distillates:

<FRACTIONS>

- High-temperature
- Medium-temperature
- Low-temperature
- Residue (ETC)

A detailed analysis is out of scope of this manuscript (see: Appendix-oil-refinery).

However, when quantitative analysis of formation of CO_{2gas} from crude oil production is made, one needs an exact process (see Appendix: crude oil). All crude oil is used after it has been treated at a suitable refinery. This entails a comprehensive separation of main components of the product. It is important to mention that some components are directly involved in combustion to CO_{2gas}. On the other hand, some are not directly converted to CO_2. In fact, some components are reverted to carbon-sink (such as plastics), and can be used in carbon-recycling (Appendix A).

It is known that over 99% of *plastics* are made from chemicals produced in oil refineries (almost 10% of crude oil is used) (El'gendy and Speight, 2002; Rosenzweig et al., 2021; Manahan, 2022). Therefore, some utilizations of plastics are carbon-sinks. In other words, almost 90% of fossil oil is left for refinery to manage distillation into various other components. Therefore, one needs to emphasize that any unit crude oil quantity is in reality distributed in everyday usage:

1. light oil fractions (such as: airplane gasoline)
2. medium fraction: gasoline
3. heavy oil (diesel)
4. precursors for different materials: medicines, plastics, agriculture, etc.

Thus if 100 million barrels of crude oil are used/produced every day, then these various fractions are in actual used in technology (e.g. gasoline/petrol/deasil/heavy fuel/tar). This quantity is related to the world population, currently ca. 8 billion. It is estimated that the world population will stabilize around 12 billion in a decade or more (Figure 1.11) (Appendix A).

Carbon dioxide as gas (CO_{2gas}) is both produced (mainly by fossil fuel combustion) and consumed (by various procedures).

This process (i.e. the production of carbon dioxide) is man-made and is currently related to the world population (e.g. food and energy demands). It is thus seen that the above process of combustion adds CO_{2gas} to air. At the same instance, it must be stressed that carbon dioxide in air is essential for the existence of life on earth.

1.6 PHOTOSYNTHESIS AND CARBON DIOXIDE CYCLE

The most significant process with regard to life and all kinds of living species on the earth is the supply/ production of food. The latter is singularly provided by the complex system which is dependent on the photosynthesis process (Birdi, 2020):

$$\text{Sun radiation} + CO_{2gas} + \text{oxygen} + \text{water} \rightarrow \text{plants/food/fisheries}$$

1.7 MASS BALANCE ANALYSES OF CARBON DIOXIDE (CO_2) AROUND EARTH

The molecule CO_{2gas} is found as gas at ordinary temperature and pressure (room temperature/1 at pressure) on earth's surface (Birdi, 2020; Appendix A). As found from geological evolutionary phenomena, it is also found that the molecule CO_2 is known to play a vital role (Appendix A). Hence

Quantitative estimate of carbon dioxide:

1. CO_{2gas} (gas) is present in atmosphere (air)
2. CO_2 (gas) is soluble in water (CO_{2aq}) (oceans; lakes; rivers)
3. Components of CO_2(soluble in water): e.g. H_2CO_3 ($CO_2 + H_2O$); $CaCO_3$ (shells; fisheries)

CARBON DIOXIDE (CO2) + SUN SHINE + WATER < PHOTOSYNTHESIS > ALL KINDS OF PLANTS/FOODS/CARBOHYDRATES

It is known that under evolution of life on earth, plant-food existed long before life appeared on earth. Plants are photosynthesis products. This means that the concentration of carbon dioxide was probably never higher than the ppm range as found currently.

The petrified plants are reported to be over half-million years old. This indicates that the CO_{2gas} equilibrium as found today has existed in pre-life era. Thus photosynthesis and GHG phenomena have existed for all these years, until today.

Each gas component in air is known to play a specific role in the life cycle on earth. This arises from the fact that chemical evolutionary equilibrium (CEE) has been stabilized over these billions of years.

For example, even though the concentration of carbon dioxide (CO_{2gas}) in air is at present very low, 400 ppm (0.04 %), it provides all the **carbonaceous food** to the existence of life. Food is obviously the most necessary product for the existence of life on earth. Actually, the photosynthesis process is opposite to the fossil fuel combustion and carbon dioxide is captured in plants. It is important to mention that all the carbon atoms in plants on earth are supplied by CO_2 in air (via photosynthesis process). Furthermore, the food is digested by all living species on earth by the metabolism process (Appendix A):

In order to analyze the climate and carbon recycling phenomena, it is important to consider the different states of material in the planet earth.

It is thus useful to consider quantitative (estimates) of carbon dioxide (CO_{2gas}) gas (which has only GHG) in the following:

EXAMPLE:

<MASS-BALACE OF CARBON DIOXIDE (with/without GHG effect)>

(estimated figures)	1 Gton = 1000 million ton)	
$CO_{2,air}$	70 Gtons	with GHG
$CO_{2,oceans}$	70.000 Gtons	without GHG

$CO_{2,gas}$ Equilibria:

$$K_{CO2,oceans} == [CO_{2,air}]/[CO_{2,oceans}]$$
$$K_{CO2,lakes} == [CO_{2,air}]/[CO_{2,lakes}]$$
$$K_{CO2,rivers} == [CO_{2,air}]/[CO_{2,rivers}]$$
$$K_{CO2,rain} == [CO_{2,air}]/[CO_{2,rain}]$$

<The magnitudes of these equilibrium constants will be expected to be different>

<Besides the component CO_{2aq}, there are additional carbonate components in oceans/lakes/rivers: carbonate salts + fisheries)>

>The $[CO_{2gas}] = [CO_{2aq}]$, equilibrium, is a pseudo-equilibrium state. This is due to a lack of sufficient mixing in the atmosphere – ocean system. It is reported that only a few hundred meters under the surface of oceans (the floors of oceans (which maybe over 5 km deep) are unperturbed. In fact, the physico-chemical data of these vast areas are unavailable in the literature.

Hence the quantities mentioned here are rough estimates.

<Other Different $CO_{2,gas}$ sources:

$CO_{2,forests}$·············...................50.000 Gtons...............without GHG

$CO_{2natural}$ gas, deposits inside earth..... Gtons................with GHG

CO_{2gas}, earthquake eruptions .... unknown ,,....... with GHG

........(Gtons =)

Furthermore, fossil fuels are estimated to be composed of approximately 70%–80% as carbon. The combustion process gives rise to carbon dioxide ($CO_{2,gas}$) and water (see). However, in all cases crude oil is primarily first treated at a refinery in order to separate it into different components, before it is used.

EXAMPLE:

PRODUCTION (man-made) OF CARBON DIOXIDE (CO_{2gsd}) FROM DIFFERENT FOSSIL FUELS

FOSSIL FUEL	CO_{2gas} (kg/m³)
COAL (carbon)	90
FUEL OIL (after refinery)	70
NATURAL GAS (CH_4)	50

Since combustion of natural gas gives rise to lower carbon dioxide emissions. Therefore, in recent decades, many energy plants are designed to use this process.

Currently, most coal/oil fired plants are converted to using natural gas for energy production and thus reducing the CO_{2gas} context. It is also found that in energy production from a coal fired player, the flue gas contains almost 10% CO_{2gas}.

The CO_{2gas} – Food – Metabolism cycle:

CO_{2gas} in air (photosynthesis) >>>> Food >>>> Metabolism (exhale CO_{2gas})

EXAMPLE:

>Carbon dioxide production and consumption phenomenon:

- Production: man-made fossil fuel combustion (released into air)
- Consumption of CO_2 from air: plant growth; food consumption; solubility in water (oceans; lakes; rivers).

1.8   CLIMATE CHANGE DUE TO MAN-MADE FOSSIL FUEL COMBUSTION POLLUTION

Mankind has been using increasingly the fossil fuels since the advent of industrial revolution (Birdi, 2020). For example, currently the magnitudes are estimated as follows:

- Crude Oil (before refinery)---100 million barrels/day
- Natural Gas---30 million eq. Of oil/day
- Coal;---30 million eq. Oil/day

Besides other climatic effects from the usage of these fossil fuels, the extent of pollution has also been observed to have increased – worldwide . Furthermore, pollution has been found to consist of various kinds:

- air pollution (gaseous – soot particles)
- oceans/lakes/rivers pollution

This shows that in the life cycle on earth all the carbonaceous component is solely provided by CO_2 in the air (EPA Handbook, 2011; Birdi, 2020). Actually, it is clear now that all these natural elements are interrelated and basic necessities for the life on earth. The characteristics of the different gases in air (atmosphere: nitrogen; oxygen; carbon dioxide (CO_2); water-vapor; dust particles) for example:

Oxygen (O_2: 21% in air) is essential for all living species (as a source of metabolic and other reactions (e.g., oxidation).

Additionally, the living species in oceans (fisheries) are completely dependent on metabolism based on oxygen intake (.). This essential food makes this gas very important for fisheries.

Furthermore, oxygen is a very essential component of all combustion reactions. In all combustion processes, when fossil fuels, wood, etc. are burned in air (containing 21% oxygen, O_2),

Fuel (consisting of carbon (and hydrogen) molecules) $+ O_{2gas}$ (gas) $= CO_{2gas}$ (gas) $+ H_2O$

All the food (photosynthesis process) for all living species (as well as all the organic molecules which mainly consist of carbon atoms), are made by natural processes (using $CO_{2gas.}$).

Most significantly, it is important to mention that the concentration of CO_{2gas} in air must be sufficient to support the need for plant growth on the earth. Therefore, one will expect that there must be a **minimum** concentration of CO_{2gas} (in air) in order to sustain life on earth (Appendix A).

Nitrogen (N_2: 78%) Even though nitrogen does not interfere directly in life cycle, it is however converted to ammonia (NH_3) after interaction with hydrogen and used as a fertilizer for growing plants (food). This means that the increasing demands for abundant food production globally (related to increase in world population) need fertilizers to assist in this demand.

Dust Particles in Atmosphere (Air Pollution): In many different phenomena on earth (both natural and man-made) so-called pollution from dust particles is observed worldwide. However, the term pollution has been applied to many other phenomena also (Appendix A).

The specific dust pollution phenomenon arises from different phenomena:

- general human activity
- general living species activities
- earthquake (erratic) eruptions
- rains/storms/hurricanes

The different gases in air (atmosphere) exhibit different physicochemical properties (as regards solubility in water and absorption of infrared light). However, there is found that an equilibrium through chemical evolution has resulted in the state of present-day state on earth.

As regards the solubility in water, it is found that the solubility of oxygen and nitrogen is very low. On the other hand, carbon dioxide is slightly soluble in water (i.e. in oceans/lakes/rivers) (IPCC, 2007; Rackley, 2017). Hence, one finds that there is a very large amount of CO_2 (as: CO_{2aq}/carbonates/shells) in oceans, which is in equilibrium with air (due to the solubility of former in water). This chemical revolutionary equilibrium was initiated billion years ago (e.g. pre-living species; pre-industrial revolution).

CARBON DIOXIDE IN AIR===(EQUILIBRIUM) ===

CARBON DIOXIDE IN OCEANS

This means that any changes in the concentration of carbon dioxide in either phase (e.g. air or oceans/lakes/rivers) will lead to a corresponding change in the other phase (Section 1.3; Appendix B).

1.8.1 CHANGE OF EQUILIBRIUM CONCENTRATION OF CARBON DIOXIDE IN AIR (CO_{2GAS}) AND IN OCEANS (CO_{2AQ})

Carbon dioxide, CO_{2gas}, in the gaseous form, is found in air. However, due to the interaction of $CO_{2,gas}$ with land and oceans (found as: $CO_{2aqueous}$) on the earth, the state of carbon dioxide is at equilibrium (pseudo) under these conditions (during the chemical evolution).

It is also important to mention the possible reason/source for CO_2 increase in air in the recent century. With the rapid increase of the global population and the industrialization of more and more countries, the demand of energy is growing almost proportional to the world population. Currently over 85% of the global energy demand is being supplied by the burning (combustion) of fossil fuels (e.g. natural gas + oil + coal). It is also expected that fossil fuels will continue to be the main source of energy in fore seeable future. The burning (combustion) of these fossil fuels releases CO2gas into the atmosphere; this disrupts the carbon (CO_2) balance of the planet which is estimated to have has been steady over hundreds of millions of years. Although anthropogenic CO2 emissions are relatively small as compared to the natural carbon fluxes, such as photosynthetic fluxes, the increased release has had obvious influences on the global climate in a very short period of time (Schimmel, 1995; IPCC, 2011; Birdi, 2020).

The estimates of CO_2 as found in different states (either emissions or as stored) are given as follows (as Gt of C: carbon; in CO_{2gas}):

- Man-made $CO_{2,gas}$ production from fossil fuel combustion (variable): 8 Gt C/year
- $CO_{2,aqueous}$ as present in biosphere and oceans: 40,000 Gt C
- $CO_{2,gas}$ in atmosphere: 780 Gt C
- $CO_{2,gas}$ as found in the exhalation (of $8 \ 10^8 = 8$ billion) by humans (there are other living species): $6 \ 10^6$ per day = Gt C/year
    (fossil fuel constitutes after crude oil has been treated in refinery)

It is obvious from these data that oceans (and lakes/rivers) are an important factor as regards the CO_2 (material balance) emission/capture technology. The $CO_{2aqueous}$ as found in the oceans is considered as being a buffer, and to absorb/release any amount of CO_{2gas} in air as needed (Appendix A). Analyses show that since the beginning of the industrial age in ca. 1750, the CO2 concentration (average) in atmosphere has increased from 280 to 410 ppm in 2017 (Figure 1.2) (IPCC, 2011; Rosenzweig et al., 2021).

The varying increase of the CO_{2gas} concentration in atmosphere influences the balance of incoming and outgoing energy (i.e. heat) in the atmosphere system (due to its GHG properties), leading to the raise of average surface temperature of earth. Thus, CO_{2gas} has often been cited as the primary anthropogenic greenhouse gas (GHG), while other GHG gases are also expected to contribute to this phenomenon. The effect of pollution in environment needs to be considered in this context.

In different industries where fossil fuels are burned (process of combustion), the flue gases are known to contain besides CO_2 some other pollutant gases (such as CO, NOx, and SO_2). In all cases, flue gases from coal/oil fired plants have been purged, for many decades, of these different pollutants. However, CO_{2gas} from these flue gases has not been captured to the same extent.

The time-scale of the data for CO_{2gas} (Figure 1.2) is obviously short as compared to geological time-scales. However, there are literature studies which do give (indirect) estimates of earth data for much longer time-scale (Appendix A).

It is also found that the seasonal concentration variation of CO_{2gas} is approximately +/– 5 ppm (i.e. Summer/winter). The concentration of CO_{2gas} drops slightly in summer due to the uptake through *photosynthesis* in plants. The latter as observed on the land (30%) area of earth. The variations of CO_{2gas} – CO_{2aq} are generally not reported in literature.

This thus indicates that approximately 5 ppm of CO_2 is captured by plants/forests. The variation of ca. ±10% arises from the carbon capture by summer months related to plant/forest growth. Carbon dioxide (CO_{2gas}), as found in air, is found to be soluble in water (Appendix A). The solubility increases (ca. three fold) at 0°C as compared to that at 30°C. In other words, in oceans (where one finds a very large amount of CO_{2aq}), concentration of CO_{2aq} would be varying with temperature (i.e. summer/winter) (IPCC, 2007).

Geological studies have shown that prior to the modern industrial human technology, the surface of the earth and atmosphere maintained a steady state with a relatively constant amount of CO_{2gas}, about 280 ppm, for several millennia before the invention of fire (the steam engines). (Appendix B). This may also indicate that the **minimum** concentration of CO_{2gas} in air would be approximately 280 ppm. Furthermore, due to the equilibrium between the two states (i.e. CO_{2gas} in air and CO_{2aq} in oceans) (equation 1.1), must be maintained. The latter is needed in order to sustain the necessary growth of foods/plants/fisheries and to support life. Over hundreds of millennia, the atmospheric CO_{2gas} concentration (and Earth's temperature) has varied naturally, due to the variations in the Earth's tilt and its orbit around the sun. The natural state of the climate for the past million years or more has been cold (glacial ice ages) with periodic warm (interglacial) periods. It is also reported that the average global (surface) equilibrium temperature has increased by about 1°C since the Industrial Revolution (Appendix B) (about 200 years ago). It is important to note that the CEE is perturbed by mankind activities, over about 200 years.

1.8.2 Temperature (Estimated) Change of the Earth (Surface) (Sun – Atmosphere – Earth)

Obviously, one cannot define any exact criteria as regards the quantity surface (equilibrium) temperature of earth. This is found from the fact that the interior of earth is fluid (lava-like) (ca. 5,000°C), while the surface crust is solid (and still changing) (temperature varies from +50°C to –50°C). In addition, the planet earth temperature is known to have changed from 5,000°C at creation to the current state.

The average temperature of the earth is primarily controlled by the sun radiation and other natural parameters, both natural and man-made. The temperature of earth which is discussed in the literature is measured at some fixed points on the surface of earth (Kemp et al., 2022; Appendix A).

SUN........ATMOSPHERE.......EARTH

This subject is out of scope of this book, but a short mention is useful for the general discussion of the climate change aspects. It is thus obvious that the amount of heat earth receives from sun is a rather complex phenomenon. Earth is also known to both receive and loose heat through different processes. It is thus useful to consider briefly the comparative quantity of sources of heat as involved in the case of climate on the earth.

All the planets in the solar system are entirely controlled by the sun (e.g. as regards different properties of the solar planets: temperature (heat input)and path (gravitational forces)

Measurements show that the average power density of solar radiation just outside the atmosphere of the earth is 1,366 W/m² (solar constant) (Birdi, 2020). It is however found that the temperature of sun (surface temperature of sun: 5,500°C) is increasing due to the chemical reactions. Also, the energy as reaching the earth is fluctuating (both as a function of time and place). The solar activity (flares and other eruptions) varies daily and yearly.

From these data one has estimated the heat energy earth receives from sun in the following process (Manuel, 2009; Birdi, 2020):

Total energy of solar radiation reaching Earth per year = 5.46 × 1,024 J = 5,460,000 EJ/year.

It is found that this quantity is very large as compared to some man-made energy sources. The annual global energy consumption (between the years 2005 and 2010) has been reported to be about 500 EJ. This amounts to ca. 0.01% (500/546,000 × 100) of the annual solar energy reaching Earth. In these calculations, the quantity average solar power on the Earth was assumed to be 1,366 W/m². However, not all solar radiation which falls on earth's atmosphere reaches the ground. About 30% of solar radiation is reflected into space. About 20% of solar radiation is absorbed by clouds and gas molecules in the air. About three quarters of the surface of earth are covered by water (oceans). The oceans therefore absorb a large part of heat from sun shine.

This phenomenon is further complicated by the fact that water shows some abnormal behavior as regards density and temperatures. However, theoretically, even if only 10% of total solar radiation is utilizable, 0.1% of which can power the entire world.

1.9 CARBON CAPTURE RECYCLING AND STORAGE (CCRS)

The **purpose of this book** is to address the current technologies available to control the carbon dioxide (CO_{2gas}) equilibrium concentration increase (due to man-made CO_{2gas} emissions). It is proposed that due to its GHG properties, carbon dioxide gas adds extra heat to the surface of earth. It must be mentioned that the GHG effect also results from other pollutants:

- methane (CH_4)
- CLOUDS (MOISTURE)
- NO_x, etc.

Fossil fuel combustion produces carbon (gaseous) which is low in concentration in flue gases (ca. 10^). This requires the need to increase the concentration of carbon by adsorption (on solids) or absorption (in suitable liquids) (Birdi, 2019, 2020) (Figure 1.13).

**(CARBON DIOXIDE
CAPTURE RECYCLING
& STORAGE)**

I

(CCRS)

**FOSSIL FUEL COMBUSTION
*CO2gas PRODUCTION)**

I

**CAPTURE
OF CO2gas**

I

I

**xRECYCLINGx
xEnhanced Oil Rcoveryx
xSTORAGE OF CO2gasx
(DEPLETED OIL/GAS RESERVOIRS;
COAL MINES; SALINE
AQUEAFIERS)**

FIGURE 1.13   Carbon (CO_2) recycling after capture (schematic).

CO_{2gas} (as gas) can be captured/sequestered by different CCRS processes (Chapter 5). This process is called carbon capture and storage or CCS (Chapter 4) (Figure 1.1).

CCRS (Carbon Capture Recycling and Storage):

Carbon dioxide (CO_2) <> Capture Process <> Storage of captured CO_2

In most current literature studies, one finds that carbon (i.e. carbon dioxide: CO_2) can be captured (from flue gases) by various methods (Bolis, et al., 1989; Hinkov et al., 2016; Rackley, 2010; Birdi, 2020) (Figure 1.14):

- GAS/*ADSORPTION* ON SOLID:
- GAS/*ABSORPTION* IN FLUID
- OTHER DIFFERENT TECHNOLOGIES TO CAPTURE CARBON (CO_2)

These literature studies are described in Chapter 2.

It is also important to mention (briefly) other related phenomena before describing the carbon capture and its surface chemistry aspects. Mankind is increasingly interested in understanding and control of the various phenomena surrounding the

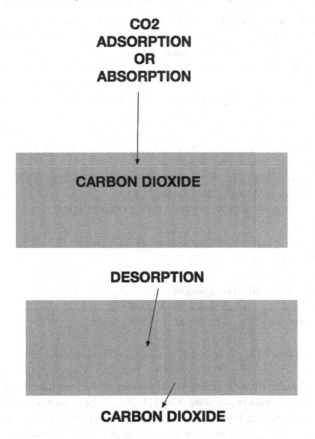

FIGURE 1.14 Carbon dioxide adsorption/absorption processes.

earth (such as: pure drinking water; clean air; average temperature of the earth). The average global temperature on earth (this term is very complex; as one might expect) is related to different energy sources/sinks, and chemical evolutionary equilibrium (CEE) between the latter.

1.9.1 CARBON RECYCLING – CARBON CAPTURE RECYCLING AND STORAGE (CCRS) (SURFACE CHEMISTRY ASPECTS)

The population of mankind has continuously interacted with natural surroundings, thus resulting in different impacts on various natural phenomena: such as environment, climate, infrastructure, economics, etc. In the present case the phenomenon related to change in the climate-temperature of earth is the main theme.

It is feasible to capture man-made CO_{2gas}, thus concentrating for further treatment.

After the capture of CO_{2gas} by the CCRS technology, the carbon (generally as liquid CO_2) needs to be stored or used in some useful application (Chapter 4). This step in CCRS has been investigated and many literature studies have been reported. The economic aspects of CCRS have also been investigated in the literature (Rackley, 2017).

The **purpose of this study** is to describe the surface chemistry aspects of carbon capture (i.e. carbon dioxide: CO_2) (Dennis et al., 2014; Birdi, 2020). The interfacial region as present in these climate aspects is described. In addition, the chemical evolutionary equilibrium (CEE) at these interfaces is found to be important factors.

The storage step may consist of different processes, depending on the nature of the application. The CCS technology thus can be described as terms means (Figure 1.4):

STEP I: Man-made Carbon dioxide (CO_{2gas}): FOSSIL FUEL COMBUSTION (BURNING)
STEP II: Man-made CO_{2gas} ==== CO_2 Capture and storage (CCS)
STEP III: Storage Methods for Captured CO_2:
STEP IV: CARBON RECYCLING AFTER CAPTURE

It is important to specify the major sources of CO_2 (where CO_2 is present as gas, in **free state**) entities:

I. atmospheric equilibrium concentration CO_{2gas}
II. man-made CO_{2gas} (from fossil fuel burning)
III. carbon dioxide (CO_{2aq}) dissolved in oceans/lakes.

This means that **free** CO_{2gas} is in equilibrium (pseudo) in I and III states (Appendix A). The state II is variable. The CCS technology can thus be applied to these two (I; II) major states where carbon dioxide (CO_2) is present in the free state. In other words, if its concentration changes in one state, it will change in the other states correspondingly (e.g. increase or decrease). Further, since carbon dioxide is soluble in water (1.45 gm/L), large quantities are thus present in the oceans/lakes/rivers (Appendix B).

Furthermore, since the CEE concentration of CO_{2gas} in atmosphere is known to be essential for the growth of plants (food/fisheries) on earth, there must be a **minimum** of its concentration in order to maintain life on earth.

CO$_2$ Equilibrium Cycle:

Minimum Concentration of CO$_{2gas}$ in Air === CO$_{2aqeous}$ in Oceans
One may estimate this minimum CO_{2gas} concentration (without any man-made CO_{2gas} emissions) to be equal to approximately 280 ppm. The latter figure is related to the average CO_2 concentration before the beginning of the industrialized revolution. The CCRS technology thus has to address the following parameters (Metz et al., 2005; IPCC, 2007, 2011; Hinkov et al., 2016; Leung et al., 2014; Rackley, 2017; Birdi, 2020):

- control/capture of man-made contribution of CO_{2gas} to air;
- reduce/control the CO_{2gas} in atmosphere

The main CCRS approach relates to:

- *Man-made CO$_2$ production (combustion of fossil fuels)*
- *Increase in CO$_2$ concentration in air*
- *Capture of man-made CO$_2$ (at the source: flue gas; directly from air)*
- Recycle of captured CO$_2$ (as liquid)
- Storage of captured CO$_2$ (as liquid)

CCRS thus relates to control and monitor of CO_{2gas} concentration in air. After this capture process, there is a need for storage or similar technology. A short specific survey of CO_{2gas} (only CO_{2gas} exhibits GHG properties) capture techniques will be given for the reader to help in future research trends and analyses (Chapter 2; Appendix B). Most of the carbon capture processes currently are mainly related to surface chemical principles (adsorption, absorption, membrane separation, cryogenic process, carbon dioxide, hydrate formation, etc.) (Section 1.2.1).

Currently, the main research on CCS is based on two main surface chemical processes: adsorption and absorption. This is the main theme of this book. The surface chemistry aspects will be delineated with the help of classical thermodynamics of adsorption theory (as described in the literature by Gibbs adsorption theory) (Chattoraj & Birdi, 1984; Adamson & Gast, 1997; Bolis, 1989; Yang, 1987; Birdi, 1999, 2009, 2016, 2017, 2020; Rackley, 2017).

However, some additional carbon capture techniques will also be briefly mentioned.

The concentration of CO_2 as found in different (major) sources (natural and man-made) varies as follows:

- in atmosphere (ca. 400 ppm = 0.04 %)
- flue gas from a coal/oil fired plant (ca. 100,000 ppm; ca. 10%)
- flue gas from a natural gas (>95% CH_4) fired plant (ca. 2%)

It means that the CCRS technology needed for these sources will be different (as related to the concentration of CO_2). Further considerations as regards the parameters regarding carbon dioxide (CO_2) variables are as follows:

- CO_{2gas} concentration in air has increased from 280 ppm to 400 ppm over a few centuries.
- CO_{2gas} is soluble in water (e.g. oceans; lakes; rivers) ($CO_{2aqueous}$ concentration has also increased in oceans)

This shows that man-made CO_{2gas} leads to an increase in its concentration both in air and in oceans/lakes/rivers (due to CO_{2gas} solubility in water as $CO_{2aqueous}$) (as: CO_{2aq}). The photosynthesis activity has also been mentioned to change. This would thus have an additional effect on the relationship between climate and plant-forest activity.

It is estimated that the man-made CO_{2gas} from fossil fuel combustion is about 25 million tons/years (increased by about 15% over the past 50 years). The cost of application of CCRS technology has been evaluated (Appendix A). This subject is out of the scope of this book and will be discussed only briefly.

It is thus seen that to define the average concentration of CO_{2gas} in the atmosphere is not quite simple. The general process of capture of CO_2 (carbon capture recycling and storage – CCRS) is (Figure 1.1; Appendix A):

CO_{2gas} (as gas) ==== CO_2 (capture process: absorbed/adsorbed/recycling/etc.)

The most plausible process generally suggested is to capture CO_{2gas} at its production site (i.e. flue gas from coal or oil or gas fired plants). Currently there are various procedures which are suggested to capture CO_{2gas} from flue gas of industrial plants (Figure 1.15). At present, there are two well established procedures of capturing gases from flue gases (Figure 1.3):

...absorption (SECTION 2.2)

(gas is captured by interaction with a suitable fluid)

... **adsorption** (SECTION 2.1)

(gas is captured by interaction with a solid)

...different other procedures for carbon capture (Rackley, 2017) (Section 2.5; Appendix A).

...Forest-plants (photosynthesis)

...Binding of CO_{2gas} to minerals (oxides; hydroxides) to form carbonates (in oceans/lakes/rivers)

...Polymer membrane gas separators (size-exclusion).

GAS ADSORPTION ON SOLID

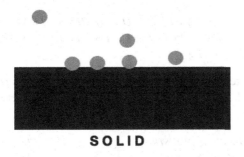

SOLID

GAS ABSORPTION IN LIQUID

LIQUID

FIGURE 1.15 Gas adsorption and absorption processes.

In the case of carbon dioxide, CO_{2gas}, capture technology, and gas absorption in suitable fluids, the gas molecules have to:

1. Interact with gas (as gas bubbles) through the surface of the fluid,
2. diffuse into the fluid (under suitable stirring (in order to reach equilibrium))

This means that the overall process will include various steps (Chapters 2 and 4). CO_2 may dissolve in the fluid. Or it may interact with some component (s) in the solution. The extensive solubility – interaction in oceans of carbon dioxide is described elsewhere herein (Appendix A and B).

1.9.2 Mass Balance of Carbon Dioxide in Different States

It is important to consider (mass balance) the **quantities** of CO_{2gas} involved in CCS technology. After the creation of earth (4.6 billion years), the surface cooling made the crust on the surface of earth.

The global CO_{2gas} emissions from different fossil fuel combustion processes are reported to be as follows (approximate data in giga-tons (Pg) CO_2/year):

NATURAL GAS (5 Pg) + CRUDE OIL (FUELS + TRANSPORT) ((10 Pg) + COAL (9 Pg). + WOOD + OTHER = TOTAL CARBON DIOXIDE EMISSIONS = 24 Pg

As reference, it is interesting to mention that the current consumption of oil globally is 100 million barrels per day. The various sources of CO_2 (wherever fossil fuels are used) emissions are:

CHEMICAL INDUSTRY	1 Pg
CEMENT INDUSTRY	1 Pg
STEEL INDUSTRY	2 Pg
TRANSPORTATION	5 Pg
ELECTRIC POWER GENERATION	10 Pg
OTHER INDUSTRIES (ETC)	5 Pg

(1 Pg $= 10^{15}$ g $= 1$ gigs ton (G t) $= 10^9$ metric tons $= 10^{12}$ kg).

1.10 MECHANISMS OF ADSORPTION AT THE GAS/SOLID INTERFACE

The gas molecules interact with the surface molecules of the solid (when in close proximity). The degree of interaction (i.e. amount of gas adsorbed per gram of solid) depends on various parameters (Chapter 2):

- surface forces between gas and solid molecules
- temperature/pressure

Experiments show that the gas – solid interaction consists of different kinds of surface forces (Adamson & Gast, 1997; Chattoraj & Birdi, 1984; IPCC, 2007; Birdi, 2010, 2014, 2017, 2020; Hinkov et al., 2016).

The *adsorption energy* is mainly dependent on the distance between molecules (i.e. between a gas molecule and the solid surface).

The simple description of gas adsorption on a solid surface can be described as follows. It is known from experiments that the surface molecules in a solid, S, are different from the molecules/atoms inside the bulk phase. A clean solid surface consists of molecules which are the same as in its bulk phase (**S**):

SYSTEM; SOLID (under vacuum)

SSSSSSSSSSSSSSSSSSSSSSS......surface layer of solid

SSSSSSSSSSSSSSSSSSSSSSSS......bulk phase of solid

SSSSSSSSSSSSSSSSSSSSSSSS......bulk phase of solid

If gas, G, is present, then the solid may adsorb with varying degree:

SYSTEM: GAS + SOLID

(MONO-LAYER ADSORPTION) (Figure 1.4)

GGGGGGGGGGGGGGGGGG......gas adsorption (monolayer)

SSSSSSSSSSSSSSSSSSSSSS.....surface layer of solid

SSSSSSSSSSSSSSSSSSSSSS.....bulk phase of solid

SSSSSSSSSSSSSSSSSSSSSS....bulk phase of solid

Or may show bi-layer gas adsorption (Figure 1.4); (adsorbed gas molecule = **G**):

GGGGGGGGGGGGGGGGGG

GGGGGGGGGGGGGGGGGG...........high gas adsorption (bi-layer)

SSSSSSSSSSSSSSSSSSSSSS.........surface layer of solid

SSSSSSSSSSSSSSSSSSSSSS.........bulk phase of solid

SSSSSSSSSSSSSSSSSSSSSS.........bulk phase of solid

Or *multilayer* adsorption:

GGGGGGGGGGGGGGGGGG

GGGGGGGGGGGGGGGGGG

GGGGGGGGGGGGGGGGGG.........extensive high gas adsorption (multi-layer)

SSSSSSSSSSSSSSSSSSSSSS.........surface layer of solid

SSSSSSSSSSSSSSSSSSSSSS.........bulk phase of solid

SSSSSSSSSSSSSSSSSSSSSS...........bulk phase of solid

This schematic drawing merely shows the various surface forces which are involved in the gas adsorption on a solid process (e.g. interactions between gas (G) and solid (S): G – G; G – S). The mechanism of gas adsorption can be determined from the experimental data. Experiments show that these models are indeed found in everyday technology. The most significant feature is that the distance between the gas molecules is reduced by a factor of 10 after adsorption. The adsorption data (i.e.

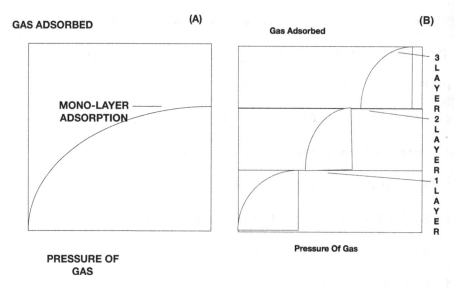

FIGURE 1.16 Gas – solid adsorption isotherms (a) monolayer and (b) multi-layer systems) (amount of gas adsorbed versus gas pressure) (idealized).

data of amount of gas adsorbed/gram of solid: pressure of gas) allow one to analyze these mechanisms (Chapters 2 and 4) (Figure 1.16).

1.10.1 Solid Surface Characteristics (Used for Gas Adsorption)

The nature of the solid (adsorbent) has a profound effect on the gas adsorption process. The basic feature of a good adsorbent (solid) is a large specific surface area (area/gram). For example, the surface area of a solid material/gram increases as the radius of the particles decreases:

<SURFACE AREA OF A SOLID VERSUS PARTICLE SIZE>

<Surface area (m²/gram) of solid particles......Increases with decrease of particle size>

<Example: Large solid sample particle:
          weight = 1 gm   Volume of solid = 1 cc (1 cm × 1 cm × 1 cm)
Shape of the solid = Cube
1 gm solid with total surface area = 1 cm² × 6 = 6 cm²
<After crushing to make cubes of size 0.10 cm:
Number of particles (in powder) = 1/(0.1³) = 1,000 particles/g.
Surface area of powder (1 gm) = 1,000 × 6 × 0.1² = 60 cm²
Increase in surface area/gram = 60/6 = 10 times.
   These surface area estimates thus show that as the number of particles per gram increases, the total surface area of solid (per gram) increases. Typical solids from everyday life:

- Talcum powder....$10\,m^2/g$
- Active charcoal... $>1,000\,m^2/g$

In the case of **porous solids**, the pores add considerable surface area for gas adsorption (Chapter 2). The gas adsorption in porous solids is found to behave as shown in the following.

BEFORE GAS (G) ADSORPTION:

PLANE

CRYSTAL SOLID.......SSSSSSSSSSS

　　　　　　　　　SSSSSSSSSSS

POROUS SOLID.....S/ / S/ /S/ /S/ /S/ /

　　　　　　S/ /S/ /S/ /S/ /S/ /

　　　　　　S/ /S/ /S/ /S/ /S/ /

(here / / indicates pores for gas adsorption)

AFTER GAS (G) ADSORPTION:

PLAIN....................GGGGGGGG

CRYSTAL SOLID.....SSSSSSSSSSS

　　　　　　　　　SSSSSSSSSSS

POROUS SOLID........**GGGGGGGGGGGG**

.............................SG/SG/S/SG/SG/SG/S/SG/SG/SG/S/

.............................S/SG/S/SG/S/SG/S/SG/S/SG/S/
(gas adsorption in pores/SG/)

The larger the surface area of the solid adsorbent (per gram of solid), the more gas molecules can be adsorbed on its surface (i.e. per gram of adsorbent). This means that the process is efficient as regards rate, cost, etc. However, the gas adsorption process in *porous* materials is complex, since it relates to the dimensions of the pores and the size of the gas. It is also found that the adsorption energy is different inside the pores, as compared to that outside the pores. Generally, this means that a good adsorbent is very porous. The characteristics of porous solids have been investigated extensively in the literature (Appendix B) (Keller & Sturadt, 2006; Yang, 2003; Birdi, 2014, 2017, 2020). This arises from the fact that pores induce extra surface for adsorption. The magnitude can vary from $100\,m^2$ to $>1,000\,m^2/g$ (Keller & Staudt, 2006; Korotcenkov, 2013; Birdi, 2003, 2016, 2020).

Porous adsorbents generally may contain pores ranging from *micropores* with pore diameters of less than 1 nm to *macropores* with diameters of >50 nm. These absorbents are found to exhibit properties of the solid surface, ranging from crystalline materials like zeolites to highly disordered materials such as activated carbons. The typical adsorbents which are used for gas adsorption are given as follows (Appendix B; Chapter 4):

- silica gel,
- activated carbon,
- zeolites,
- metal organic frameworks,
- ordered mesoporous materials,
- and carbon nanotubes.

The specific area of an adsorbent is the surface area available for adsorption per gram of the adsorbent. Special procedures (gas adsorption experiments) are used to determine this quantity for a given solid. This quantity is a specific physical property of the solid surface. In order to understand the gas adsorption process, this quantity is needed.

GAS PHASE

G   G   G   G

– SURFACE OF SOLID
In order to estimate the area/molecule of the adsorbed gas, one needs to determine:

- amount of gas adsorbed
- surface area of solid material/gram

If one knows the gas molecule dimensions (from other measurements), then one can estimate the packing of the adsorbed layer(s) (Chapter 2). This means that one needs the following data:

- amount gas adsorbed/surface area of solid/gram
- area/molecule of gas

It is also obvious that in all gas – solid systems, one needs the information about the surface area/gram of the adsorbent, in order to make physical analyses. The mechanism of adsorption also requires the information about the quantity amount of gas adsorbed/gram of solid as the sorbent (as a function of temperature and pressure) (Figure 1.5) (Chapter 4).

1.10.2 GAS/SOLID ADSORPTION ISOTHERMS

One may expect that the process of gas adsorption on a solid may be a relatively simple process. However, experiments show that it is a somewhat complex system (Chapters 2 and 4). The gas adsorption process is generally investigated by measuring

**Gas adsorbed
mmol/gm**

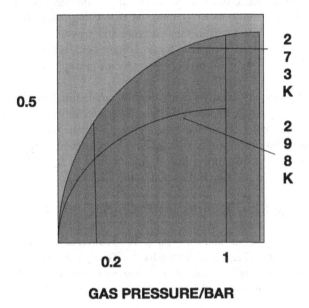

GAS PRESSURE/BAR

FIGURE 1.17 Gas adsorption on solid (effect of temperature): amount of gas adsorbed (mmol/gm) versus gas pressure.

the amount of gas adsorbed per surface area of the solid material. The amount of gas adsorbed on a solid is measured as a function of pressure (Figure 1.5). The effect of temperature is also measured in most instances (Figure 1.17).

In general:

- gas adsorption on solid surface increases with higher pressures.
- gas adsorption on solid surface increases at lower temperature.

Typical CO_{2gas} adsorption data on activated carbon (Suzuki, 1991):

[temperature----------PRESSURE-->...0.2 bar............................1.0 bar]

273K..0.3 mmol/gm...............0.7 mmol/gm

298K..0.2 mmol/gm...............0.5 mmol/gm

A similar kind of adsorption behavior is found for various other gas – solid adsorption systems (Chapters 2 and 4) (Birdi, 2020).

1.10.3 GAS ABSORPTION IN FLUIDS TECHNOLOGY

This process is described as the gas absorption in a fluid (Figure 1.3). The gas molecule may be soluble or interact with the components in the fluid. The removal of a gas from flue gas by bubbling (scrubber) through a fluid has been used for many decades (Birdi, 2020).

The solubility characteristics in water of carbon dioxide gas (CO_{2gas}) are found to be complex (Appendix A). It is reported that CO_{2gas} solubility is 900 cc per 1,000 cc of water at STP (standard temperature and pressure: 20°C; 1 atmosphere). However, CO_{2gas} dissolves in water to form carbonic acid, H_2CO_3. The latter dissociates into $H^+ + HCO_3^-$. The equilibria of these species in water have been investigated (Appendix A).

For example, aqueous solutions (with mono-ethanolamine) have been used to capture CO_2 from flue gases (Chapter 4). In these CCRS technologies, aqueous solutions of various amines (with basic-properties:, e.g. mono-ethanolamines) have been used. These aqueous solutions of amines preferably strongly bind CO_{2gas} at low temperatures. However, at higher temperatures it is found that CO_{2aq} is desorbed as CO_{2gas} from the solution. The reaction is depicted in the following:

$$CO_{2gas} + 2\ HOCH_2CH_2NH_2 \leftrightarrow HOCH_2CH_2NH_3^+ + HOCH_2CH_2NHCO^{2-} \quad (1.1)$$

The recovered CO_{2gas} is around 90% pure. The recovered gas is stored or used in different applications (Chapter 4; Appendix A). The gas absorption method for carbon capture has been studied extensively in the current literature. A short literature survey is given later in the text (Chapter 2).

1.10.4 Carbon Dioxide Capture by Different Processes

Besides adsorption and absorption CCRS technologies of capturing CO_{2gas}, there are also other phenomena which can be used to capture CO_{2gas}. For example:

- The plants (of all kinds) grow through a process called photosynthesis. The photosynthesis which all plants use to grow, captures CO_{2gas} from air (Pessarakli, 2001). It has been suggested that increased plantation of forests/trees thus assists in very significant capture CO_{2gas}.
- Fisheries also depend on photosynthesis + carbon capture.

EXAMPLE

<CARBON CAPTURE ESTIMATES FOR TREES (PHOTOSYNTHESIS)

    <Average carbon captured in trees = 8 C kg/m^2
    <Average rate for trees = 0.3 kg C/m^2/year
    <OTHER CARBON – TYPICAL TREES-CO2 DATA: For USA>
    <Total storage in trees (estimated) = 640 million tons C

- Formation of inorganic salts from CO_{2gas}: calcium carbonate, magnesium carbonate, etc.
- Capturing of CO_{2gas} from a source such as any industrial plant (flue gas) or directly from air (Dubey et al., 2002).
- Carbon dioxide (CO_{2gas}) is known to be soluble in water (Chapter 4). Hence, a very large quantity of CO_{2gas} is present in oceans/lakes/rivers. At normal temperature/pressure, solubility of CO_2 is about 1.45 g/L.

<SOLUBILITY OF CO_{2gas} IN WATER/AQUEOUS (CO_{2aq}) (OCEANS/LAKES/RIVERS/RAIN DROPS):

Another very significant chemical process is the equilibrium between CO_{2gas} in air and CO_{2aq} in oceans. This arises from the fact that:

- earth is covered by about 70% oceans (plus lakes: rivers).
- CO_{2gas} is soluble in water (1.45 g/L)
- CO_{2gas} in atmosphere interacts with the surface of ocean/lake/river interface

The state of CO_{2gas} in equilibrium with oceans/lakes/rivers is different from that one finds over the land. Further, this characteristic, i.e. the equilibrium between the CO_{2gas} in air and water, is instantaneous. The chemical equilibrium between the concentration of CO_2 in the oceans (as: $[CO_{2aq}]$) and atmosphere can be described as (Figure 1.5):

$$\mu_{CO2,air} = \mu_{CO2aq} \qquad (1.2)$$

The equilibrium constant (ideal), K_{CO_2}, is:

$$K_{CO_2} = [CO_{2gas}]/[CO_{2aq}] \qquad (1.3)$$

{Activities should be used in case of real example: Quantities Activity is seldom equal to concentration} (Chattoraj & Birdi, 1984)

where $[CO_{2,gas}]$ and $[CO_{2,aq}]$ are concentrations in air and oceans, respectively (Appendix A). Further, only CO_{2gas} exhibits GHG properties (Figure 1.18).

$\mu_{co2,air}$ and $\mu_{co2,ocean}$ are the chemical potentials of CO_2 in air (atmosphere) and ocean, respectively (Appendix A). This means that if CO_2 concentration changes in either of the phase, the equilibrium changes its concentration in the other phase. Thus if concentration of CO_2 increases in air, then it will induce an increase in concentration in ocean/lakes (as follows from the chemical potential equilibrium equation 1.1). In other words, currently there is 410 ppm CO_2 in air, which is in equilibrium with $CO_{2aqeous}$ in the oceans. If the concentration of CO_{2gas} is reduced (or changes) to 300 ppm in the air, then some equivalent amount of CO_{2aq} will be released from the oceans, to satisfy the chemical equilibrium state (equation 1.1).

Most significant is that this equilibrium is instantaneous, and the amounts involved (in the oceans) are also very large in comparison with man-made CO_2 from fossil fuel

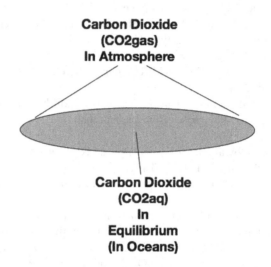

ATMOSPHERE - OCEAN
INTERFACE
(70% Of Earth Surface)

Carbon Dioxide
(CO2gas)
In Atmosphere

Carbon Dioxide
(CO2aq)
In
Equilibrium
(In Oceans)

[] Largest Gas - Liquid Interface []

FIGURE 1.18   Carbon dioxide equilibrium between air <> oceans/lakes/rivers.

combustion (Dennis et al., 2014). However, CO_{2aq}, as found dissolved in the oceans (lakes/rivers), does not contribute to the greenhouse phenomena. It is reported that ca. 95 Pg of carbon (CO_{2gas}) is transported across the atmosphere –.ocean *interface*. The carbon dioxide equilibrium at the ocean surfaces is very fast. It is estimated that about one third of anthropogenic CO_{2gas} is absorbed by the oceans (equation 1.2) (Kemp et al., 2022; Birdi, 2019; IPCC, 2007).

In other words, the oceans can absorb and store considerable amounts of CO_{2gas} (due to their extreme volume) (Birdi, 2020; Kemp et al., 2022). This amount in oceans will be more than (>50 times as the heat absorbed on the sun = atmosphere – land area (Kemp et al., 2022). Considering that in this system, one has pseudo-equilibrium.

Furthermore, due to the anomalous water physical properties, the CO_2 – water (i.e. oceans) equilibrium is not as simple as one may assume as a first consideration. Water is the only fluid which exhibits a maximum density (at 4°C) (Birdi, 2014, 2017). This gives rise to some exceptional features. For instance, as temperature reaches 4°C, then the water at surface flows to the bottom of ocean. This means that the circulation in oceans due to temperatures around 4°C will also contribute to CO_2

cycle since the latter is soluble in water. The tidal phenomenon, in oceans, is related to the gravity interaction with the moon. These tidal phenomena give rise to turbulence and mixing in the upper layers of oceans.

1.10.5 APPLICATION OF CARBON CAPTURE AND RECYCLING (CCR)/STORAGE (CCRS) ON CONTROL OF CARBON DIOXIDE CONCENTRATION CHANGING AIR

Mankind has become dependent on energy as produced from fossil fuels. However, the inherent trait in mankind is also the quality of innovation technology. This property has led to innovations which decrease the energy needs, without any significant compromise in daily needs. This dependence, however, adds carbon dioxide to the atmosphere and its GHG effect. Currently, there are various technologies in progress to apply CCR/CCRS to mitigate the GHG effect from carbon. This mitigation is already observed wherever energy is produced from any non-fossil fuel source (worldwide). Indirectly, the latter also decreases the man-made pollution in different aspects. Man-made carbon dioxide (CO_{2gas}) is both a GHG and a pollutant.

In previous reports, it has been determined that (Inter-governmental Panel on Climate Change (IPCC) CO2gas emissions (man-made) could be reduced by 80%–90% for a modern power plant that is equipped with suitable carbon dioxide capture and recycling (CCR) or storage (CCRS) technologies (Figure 1.1). The applications of both CCR and CCRS technologies will also complement other crucial strategies, such as changing to less carbon content fuels, improving energy efficiency, and phasing in the use of renewable energy resources (Rackley, 2017; Birdi, 2020). In a different context, the atomic energy, hydro-energy, solar energy, and wind energy are almost CO_2 free technologies.

1.10.5.1 Direct Capture of CO2gas from Air (Current Concentration of 420 ppm: 0.042%)

It is reported that the current GHG effect arising from $CO_{2,gas}$ in atmosphere is high. This is generally attributed to the increasing combustion rate of fossil fuels since the industrial revolution (Appendix C).

At worldwide effort attempts, with some success agreements are currently signed under COP meetings (COP:). However, in this context, one needs to mitigate the current concentration of CO_{2gas}. This arises from the fact that this is higher than normal.

The technology of *direct air capture* refers to the process of removing CO_{2gas} directly from the ambient air (with concentration 400 ppm) (as opposed to from point sources). As mentioned elsewhere, there is a minimum concentration of CO_{2gas}, ca. 280 ppm, and hence current CO_2 concentration is ca. 100 ppm in excess (400–280 ppm). It is obvious that due to very low concentration of CO_{2gas} in air, the methods have to be modified accordingly adapted. Currently, there are various technologies which are developing and aimed at the mitigation of reduction/control of CO_{2gas} in air (Keith et al., 2018).

1.10.5.2   Geological Storage/Recycling of Captured Carbon (CO2gas) (CCRS)

After carbon dioxide (CO_{2gas}) is captured (generally about 90% pure) (CCRS-process), it needs to be stored in some safe process (or used in some suitable application). This technology is about handling of captured CO_2. The captured CO_{2gas} has been stored under varying procedures. This procedure has also been called as *geo-sequestration*; this method involves injecting carbon dioxide, generally in liquid state into suitable underground geological formations (Appendix B):

- old coal/oil spent reservoirs
- other suitable underground storage structures
- oceans as carbonates ( )

It is expected that various physical (e.g., highly impermeable cap-rock) and geochemical trapping mechanisms would prevent the CO_{2gas} from escaping to the surface (Leung et al., 2014).

In the case of oil recovery from reservoirs, it is known to be a multistep process. Most of the recovery is primarily based on original pressure in the reservoir (producing around 20% of oil in place (Birdi, 2017). CO_2 has been used in oil recovery processes. It is sometimes injected into declining oil fields to increase oil recovery. It is found that about a few hundred million metric tons of CO_{2gas} are worldwide injected annually into these oil reservoirs.

1.10.6   FLUE GASES AND POLLUTION CONTROL

The content of CO_{2gas} in flue gases is different in various technologies. For example, flue gas from coal fired plants contains 9%–14% carbon dioxide (besides other gases). The natural gas (mostly methane: CH_4) fired plants have flue gas with ca. 4% carbon dioxide content. The exhaust from gasoline/diesel cars will also vary.

It is estimated that without the application of carbon dioxide control (such as CCRS technology or other) technology, CO_{2gas} concentration in air would increase to 650–1,550 ppm by 2030 (while global GHG increasing 25%–90% level of 2000) (Li et al., 2011; IPCC, 2013, 2018; Phelps et al., 2015). It is also generally suggested that while different pollutants from flue gas are captured/removed, CO_{2gas} should also be treated the same way. In a different context, the technology of cleaning flue gas (of different pollutant gases: CO, NO_x, SO_2) is very advanced at the present stage. Therefore, the application of carbon capture in some technology approach has been suggested. Of course, the quantities involved are much different, since the concentration of CO_{2gas} is much higher than that of the other gas pollutants (Appendix A).

EXAMPLE:

- Many living species (human; animals; insects) exhibit metabolism where CO_{2gas} is very crucial. The interaction of mankind-food-photosynthesis is one of these phenomena (Appendix A). In the case of humanoids and the quantity of carbon dioxide, one may estimate this by a simple method.
- Estimation of the amount of CO_{2gas} exhaled by 8 billion people (current world population):

  - Inhaled air $= 0.04$ $CO_{2gas}/20\%$ O_2
  - Exhaled air $= 4\%$ $CO_{2gas}/16\%$ O_2

2 Carbon Gas (Recycling: Adsorption/Absorption and Essentials)

2.1 INTRODUCTION

During the evolution process pertaining to mankind, fire (from wood burning: man-made addition of carbon dioxide to atmosphere) was discovered some thousands of years ago. Later, about two centuries ago, some parts in the world discovered large deposits of fossil fuels: e.g. crude oil and gas. However, the extensive usage of fossil fuels only took place, worldwide, about a century ago. This most likely relies to the difference in developed and developing - countries. Western developed countries, even today, use more energy, per capita, than the developing (Kamine and Chiangmai, 1965; Lomborg, 2022; Kemp et al., 2022). However during the evolutionary chemically equilibrium (CEE), the man-made activities perturbed this equilibrium phenomenon. The CEE phenomenon has taken place over about 4–5 billion years. This time-scale is not accountable in any man-made climate model attempt.

However, due to different technical and man-made innovations and other social/geo-political developments, the energy usage per capita is being monitored at all levels. For example, car mileage per unit gasoline/diesel has decreased by a factor of <4> over the past few decades (Havercroft et al., 2011; Rosenzweig et al., 2021; Gates, 2021; Lomborg, 2022). In addition, the innovation in electronic industry, in general, has advanced from high voltage tubes/transistors/microelectronics. This innovation is currently dynamic, and goals for better and efficient energy-sources are being developed.

Currently, usage of fossil fuels (coal-crude oil-natural gas) worldwide is found to be changing as:

- western world: −3%
- developing countries (China, India, etc.): +2%

These estimated observations will thus suggest that the carbon dioxide (CO_{2gas}) emission in western countries is decreasing and lowering (per capita) the emissions/pollution also. In fact, this trend has been observed already in various innovations, e.g. in applications where energy could be produced by non-fossil sources.

Simultaneously, the innovative methods to carbon recycle/capture are being developed.

The *current equilibrium concentration of carbon dioxide* (CO_{2gas}) in atmosphere is 420 ppm (0.042%). It is too low for any significant process.

DOI: 10.1201/9781003300250-2

However, all living species (especially mankind) have established metabolism based on lung function and body temperature, on the composition of air (Chapter 2; Appendix A) (Birdi, 2020).

EXAMPLE

$<CO_{2gas}$->Food/Fisheries (Photosynthesis)->
Mankind (other living species) -> Metabolism ->
CO_{2gas}

This is one of the most essential CO_{2gas} recycles for the existence of mankind and other living species).

Furthermore, with the current world population at about 8 billion, the quantity of carbon dioxide recycled is very significant. The GHG phenomenon in this daily recycling is nonlinear.

EXAMPLE

- >Assuming that each person intakes 2 kg carbohydrates food, per day.
- >Assuming that each person exhales about 1 kg of carbon dioxide/day
- >This corresponds to 8 billion kg of CO_{2gas}/day

Since food consumed is produced by photosynthesis (CO_{2gas},), this amount is *recycled* daily.

Therefore, carbon capture recycling and storage (CCRS) processes are being used/developing worldwide. These CCRS processes will control/reduce/mitigate the increasing concentration of carbon dioxide in atmosphere. The relevant surface chemistry aspects are discussed in this chapter.

Currently, fossil fuels (e.g. crude oil/ natural gas (after being treated at a refinery)/natural gas (CH_4)) were discovered almost two centuries ago. Most oil reservoirs may also consist of natural gas- crude oil at high pressure and temperature. This leads to treatment, i.e. refinery, before usage in technologies. At the present stage, about 300 million barrels equivalent to oil are used worldwide. This quantity is steadily known to increase along with the increasing production of energy and other needs for humans (e.g. transport housing, food/fisheries, health/hospitals/medicine, and research). This has a direct relationship with the humanoid population increase (health, life style, and span). However, due to innovations concerning the combustion of fossils on climate, some technologies are currently replacing fossil fuels with other sources for energy production. In fact, some countries are already carbon-neutral. This trend is mostly evident in the western developed countries (Figure 2.1).

Thus it is found that:

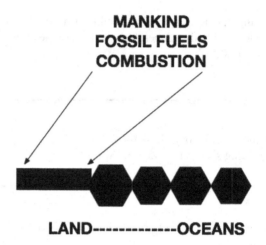

FIGURE 2.1   Mankind fossil fuel combustion mainly on land on earth.

- mankind usage of fossil fuels produces flue gas (with about 10% CO_{2gas})
- flue gas from man-made industry is mainly produced on 30% earth land-surface
- the diffusion of CO_{2gas} from land area to oceans (70%) is entirely determined by natural processes (and pseudo-equilibrium)
- man-made fossil fuel combustion related pollution (on land) does not diffuse completely towards ocean-atmosphere.

These studies show that the carbon dioxide (CO_{2gas}) emissions are very unevenly distributed. Hence any effect on environment (or climate) will be unpredictable. Therefore, most of these climate models are estimates. The application of surface chemistry principles has proven to mitigate the increasing carbon impact on environment.

Additionally, the mankind innovations have had multiple impacts on this matter. These different factors are erratic due its inherent origin. Most of the characteristics are as follows:

- supply of fossil fuels is finite and declining, and thus the CO_{2gas}-production.
- fossil fuel supplies are dependent on geo-political constraints
- important man-made innovations reduce fossil fuels (non-fossil fuels for energy: wind/solar/nuclear/others).

Change of carbon dioxide (man-made) added to atmosphere versus world population (estimated) (see: climate models: Cox, 1975;). Most of the climate related models predict that, around 2050, most of the different parameters will either level off, or show a decrease.

Most energy-climate models predict the type of result as depicted in Figure 2.2.

Fossil fuels have been used by mankind for the production of energy since the industrial revolution.

DIFFERENT SUN
RADIATION HEAT
INTERFACES

(a) SUN---INTERFACE---ATMOSPHERE
.....................I
(b) SUN---INTERFACE------CLEAR DAY
.....................I
(c) SUN------INTERFACE------CLOUDY
.....................I
(d) SUN------INTERFACE-------STORM

FIGURE 2.2   Different variable interfaces: (a) Sun – atmosphere; (b) atmosphere – earth (land); (c) atmosphere – earth (oceans).

This has led to increasing fossil fuel combustion. In addition, the concentration of carbon dioxide in environment (atmosphere) has been found to have increased.

These technologies are known to have many surface chemistry aspects (Birdi, 2020). For instance, sun heat radiation interacts with atmosphere interface. Following this stage, heat is propagated to the surface of earth (Figure 2.2). These different and varying paths have complex consequences on climate (Salty, 2012; Kemp et al., 2022; Birdi, 2020).

The latter is the subject of this study. In addition, innovative technologies have been applied to control and reduce this man-made carbon input into environment. The latter is the carbon-capture-recycling-storage (CCRS) innovation.

The present subject matter related to the order to explain the carbon recycling, the dynamic of the one needs a brief description of the sun – earth system will be useful. These significant interfaces are highlighted and discussed in relation to climate change.

EXAMPLE

<CARBON CAPTURE (Different Methods) – RECYCLING – STORAGE (CCRS)
<MAN-MADE TECHNOLOGIES>
<$CO_{2,GAS}$ EQUILIBRIUM CONCENTRATION IN AIR (400PPM: 0.04 %) – CAPTURE --> CO_2 (LIQUID STATE) (>90%)
<$CO_{2,GAS}$ IN FLUE GASES (10,000 ppm: CA.10%) – CAPTURE --> CO_2 (LIQUID) (>90%)
Furthermore, one finds natural carbon dioxide (CO_{2gas}) capture phenomena:

<NATURAL CO_{2gas} CAPTURING PHENOMENA><Interface>
<RAIN (WATER DROPS) BY CO_{2gas} SOLUBILITY><Interface>

<SOLUBILITY OF CO_{2gas} IN WATER (as $CO_{2aqeous}$: AQUEOUS SOLUTIONS) (OCEANS/LAKES/RIVERS)><Interface>
<FISHERIES INDUSTRIAL FOOD (capture and recycling of CO_{2gas} +> Shells/ Fisheries)

It is recognized that CO_{2gas}: sinks (carbon sinks) are also found in various natural phenomena:

- rain drops -> (CO_{2gas} solubility in water (rain drops))
- storms/hurricanes (rain drops)>
- oceans: the largest carbon sink (pseudo equilibrium)
- plants: foods
- fisheries: shells

The interface between atmosphere and rain drops is very large and significant. These are known from the calcification phenomenon on environment.

These different phenomena (natural carbon sinks) arise from the fact of CO_{2gas} solubility in water.

Furthermore, these phenomena are known to contribute significantly. Especially, rains are known to have inflicted a strong calcification (due to CO_{2aq}) effect on environment. This has had a strong effect on different monument's worldwide (such as: pyramids; other historical monuments)).

The planet earth is surrounded by atmosphere (gaseous). Atmosphere is composed of different gases (nitrogen > oxygen > carbon dioxide > water-vapor > others) (Birdi, 2020). The gas molecules in the atmosphere are attracted to the earth via van der Waals forces (Chapter 2).

All gases, as found in the atmosphere, exhibit different characteristics. However, **carbon dioxide** gas exhibits some characteristics which are found to have very useful effects on various parameters.

EXAMPLE

<CEE CONCENTRATION OF CARBON DIOXIDE GAS IN ATMOSPHERE – RELATION TO MANKIND>

1. PHOTOSYNTHESIS (PLANTS-FOODS-FISHERIES)
2. METABOLISM (FOOD) BY ALL LIVING SPECIES
3. SOLUBILITY IN WATER (OCEANS/LAKES/RIVERS)(FISHERIES) (rain)

4. CLIMATE AND GHG IN ATMOSPHERE
5. carbon (CO_{2gas}) mass balance in atmosphere + oceans + forests + plants system
6. GHG Phenomena have existed for billions of years (from pre-living species to today)

The GHG process has thus existed in the environment, for billions of years, since photosynthesis was observed on earth. This shows that, as regards CEE, the sun-atmosphere-earth system has reached an (pseudo) equilibrium (Figure 2.3).

This conjecture has a profound impact on the various chemical evolutionary equilibria:

- CEE
- Interfaces and surface chemistry aspects.

Earth is the only solar planet which has supported living species (mankind and other living species) over a large geological period (Birdi, 2020; Calvin, 1969). This arises

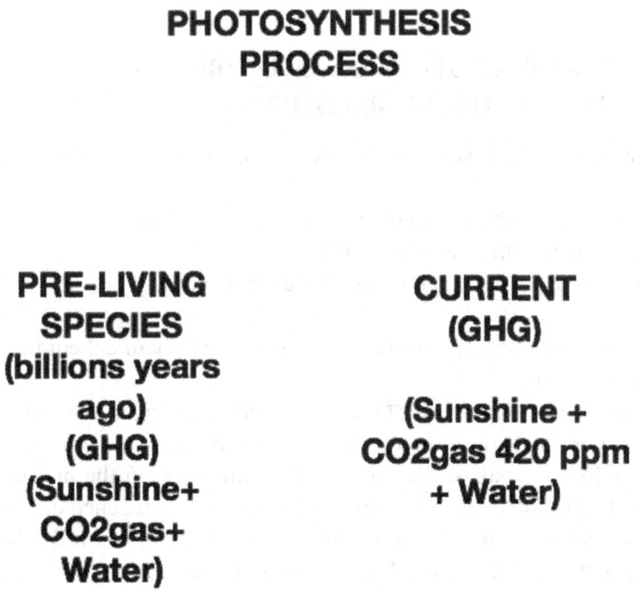

PHOTOSYNTHESIS PROCESS

PRE-LIVING SPECIES (billions years ago) (GHG) (Sunshine+ CO2gas+ Water)

CURRENT (GHG) (Sunshine + CO2gas 420 ppm + Water)

FIGURE 2.3 Earth and gaseous atmosphere.

from a combination of different natural evolutionary processes (which are out of context) (such as photosynthesis):

*   EARTH-ATMOSPHERE>
*   EVOLUTIONARY SINCE 4.6 BILLION YEARS>
*   atmosphere
*   cooling of the earth surface (formation of the crust) (formation of water; photosynthesis (growing plants with carbon dioxide + water + oxygen))
*   photosynthesis phenomenon based on: sun/carbon dioxide/water: plants/ food
*   carbon atmosphere + ocean + plant mass balance (carbon chemical equilibrium)

The evolutionary process of *photosynthesis* is known to have existed since pre-living species era (Calvin, 1969). Thus carbon dioxide was available for photosynthesis and plant growth during the chemical evolutionary equilibrium (CEE) (Section 2.2).

In evolutionary terms, this means that currently photosynthesis has reached a pseudo-equilibria and has supported the various earth systems:

<0> creation of planet earth: hot lava-like/pre-water/pre-living species
<1> pre-living species – photosynthesis/plants/forests/food---GHG
<2> current/photosynthesis/plants-food-fisheries/living species---GHG

2.2 CHEMICAL EVOLUTIONARY EQUILIBRIUM (CEE) (FOSSIL FUEL COMBUSTION)

Environment on earth (e.g. sun-atmosphere-earth) has evolved on the basis of:

*   interfaces (atmosphere – land; atmosphere – oceans)
*   chemical evolutionary equilibria (CEE)
*   synthesis of all carbonaceous substances from CO_{2gas}

All chemical phenomena have reached an evolutionary chemical equilibrium (**CE)E** (pseudo-equilibrium).

All *carbonaceous* (organic) substances as found on earth are all made from carbon dioxide (in gaseous state: at room temperature and pressure) as found in atmosphere and elsewhere (oceans, lakes, and rivers). It is important, in the present context, to mention that the composition of gasses in atmosphere has reached an (pseudo) equilibrium. This has taken a time span over the evolution. In other words, the living species, especially mankind, has reached an (pseudo) equilibrium. In this context, the evolutionary equilibrium has been the main parameter which needs to be mentioned (Figure 2.4).

CARBON DIOXIDE (CO2gas) ALL CARBONACEOUS PRODUCTS

Land
Food

OCEANS
+
LAKES
+
RIVERS
(FISHERIES)

FIGURE 2.4 All carbonaceous substances are chemical products of carbon dioxide, CO_{2gas}.

EXAMPLE

<ALL CARBONACEOUS SUBSTANCES ON EARTH ARE SYNTHESIZED VIA PHOTOSYNTHESI – SUNSHINE – CARBON DIOXIDE (CO_{2GAS})_WATER

EXAMPLE

<CHEMICAL EVOLUTIONARY EQUILIBRIUM (CEE) >

<PHOTOSYNTHESIS HAS EXISTED FROM PRE-LIVING SPECIES (Billions years Ago)   UNTIL TODAY
>MANKIND & EQUILIBRIUM BETWEEN ATMOSPHERE
>FOOD and Photosynthesis
>ALL CARBONECEOUS SUBSTANCES AND ATMOSPHERE

The chemical equilibrium (as observed today) of different substances during the evolutionary age of earth has taken place from the creation of the earth until today (Calvin, 1969; Birdi, 2020). The chemical equilibrium is of different characteristics:

1. it is pseudo-equilibrium;
2. the equilibrium adapts to the changing of environments.
3. the magnitude of CEE is much larger than any man-made perturbation.
4. CEE has been adjusting to varying environments and thus has reached a state of suitable adjustments with its surroundings
5. Industrial revolution initiated about 200 years ago:

Worldwide population has grown from about 1 billion to 8 billion today; average life-span has grown from 40 years to over 80 years.

The occurrence of pseudo-equilibria indicates that any climate model (prediction) will be subject to a lack of high degree of accuracy.

Surface chemistry principles at any interface, e.g.:

- gas – liquid (atmosphere – oceans)
- gas – solid

indicate that both phases are at equilibrium (Figure 2.5).

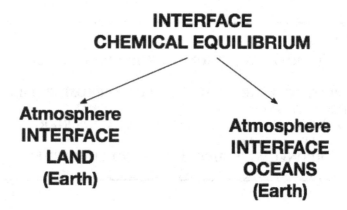

CHEMICAL EVOLUTIONARY EQUILIBRIUM (CEE) INTERFACES IN SUN-ATMOSPHERE-EARTH

INTERFACE CHEMICAL EQUILIBRIUM

Atmosphere INTERFACE LAND (Earth)

Atmosphere INTERFACE OCEANS (Earth)

FIGURE 2.5 Chemical evolutionary equilibrium (CEE) as found in the **sun** – *atmosphere* – **earth system** (schematic): different interfaces.

In this context, the significance interfaces one has to consider:

- CO_{2gas} (atmosphere) - land
- CO_{2gas} (atmosphere) - oceans (CO_{2aq})(e.g. carbonates/shells)

The latter two quantities are significantly different interfaces.

Oceans cover about 70%, and land 30% of earth surface. The carbon – carbonate – shell – fishery cycle from the interface of atmosphere – oceans is known to have played a significant role.

Furthermore, oceans can reach, on average, over 5 km depths in many parts of the world. This has been found to show that most of the heat absorbed by the earth takes place on the oceans.

It is useful to briefly consider the evolutionary age of different species surrounding the earth.

The age of earth is estimated as about 4 billion years (Birdi, 2020; Calvin, 1969) (Appendix A).

However, mankind is the only living species which has been able to interact with natural surroundings (such as: climate –pollution/changes and impacts in other manners).

This is known to relate to various phenomena: control of ozone layer; air pollution; etc. In this context, climate is also reported to change due to change in pollution.

The environmental phenomenon of <Pollution> is known to arise from both natural and man-made activities (such as industrial technologies and food/fisheries production).

Further, the effect of man-made industrial activities (since the industrial revolution) has obviously contributed to the climate (as regards: increasing air-pollution; oceans-pollution; other factors).

EXAMPLE

AGE OF EARTH AND LIVING SPECIES (Mankind/other species)
  TWO STAGES:

>STAGE I:
Beginning: Earth (creation) as a hot lava-like hot ball: 4 billion years
    >STAGE I. A:
            EARTH COOLING CRUST FORMATION: PRE-WATER/
PRE-PHOTOSYNTHESIS

>>STAGE II:
Today: EARTH~CRUST (COLD)(ABSENCE OF WATER/OCEANS/LAKES/
RIVERS/ANTARCTIC)
EARTH~ INTERIOR (HOT-LAVA-LIKE)(see: STAGE I)

<EARTHQUAKES: EARTH IS STILL COOLING WITH RESPECT TO
EVOLUTION DUE TO HOT LAVA IN EARTHQUAKE ERUPTION TO COOLER
SURFACE OF EARTH
    <EARTHQUAKE TRANSFERS HEAT FROM INTERIOR OF EARTH TO
THE COOLER SURFACE AND HEAT IS EMITTED TO ENVIRONMENT

This is obviously a very simplified estimate, due to the lack of any direct evidence data (Calvin, 1969)

Furthermore, the geological history and developments between stages I and II have remained dubious for obvious reasons (Calvin, 1969; Birdi, 2020).

In the present context, the global concentration of carbon dioxide (in air (atmosphere) + oceans + lakes + rivers) is found to be comparatively increasing (in relation to increasing fossil fuel combustion over the past few decades since the industrial revolution) over the past century (Figure 2.1). For example in atmosphere (air): currently, the equilibrium concentration is around 0.042% (420 ppm (parts per million)). Surprisingly, even though the concentration of CO_{2gas} is low, elsewhere, CO_2 still provides all the carbonaceous material needed for foods (plants + fisheries) for the existence of life on earth (including: about 8 billion people). In this context, one must mention that the equivalent of addition of CO_2 is found almost 50 times more in the oceans around the earth.

EXAMPLE:

<EFFECT ON CLIMATE (MAN-MADE)
    >MAN-MADE CHANGES AROUND ENVIRONMENT

    >**Pre-Industrial Age**-----FIRE---POLLUTION
    >**Pro-industrial age**-----FOSSIL FUEL COMBUSTION: POLLUTION/
       PRODUCTION OF DIFFERENT GAS EMISSIONS (CO; CO_2; SO; SO_3;
       NO_x')

There exists a strong connection between the criteria pollution and climate and addition of different pollutant gases (Birdi, 2020) to the environment. Most significant ones of these gases are as follows:

- NO_x
- CO
- CO_2
- SO_3

$CO_{2,gas}$ in air (in chemical equilibrium with $CO_{2,aq}$ in water (e.g. oceans-rivers-lakes) converted to food (food produced on land (corn/rice/etc.) plus food from oceans/lakes (fish + oysters + other fisheries) for mankind (alone) (for simplicity only mankind is considered; though one must consider food also produced for animals/fish/etc.).

One may consider a simple estimation based on following criteria:

POPULATION AND GREENHOUSE GAS (GHG) ESTIMATION:
Number of people (worldwide) = 8×10^9 people
Average food (estimated) per person (daily) = 1 kg/person/daily
Food needed for all mankind/day = $8 \times 10^9 \times 1 = 8 \times 10^9$ kg/day
= 8×10^6 tons/day

Assuming that all food is metabolized, and CO_2 (carbon dioxide) exhaled is 50% of food?
Carbon (carbon dioxide) produced by humans from food (metabolism)
= $8 \times 10^6 \times 0.5$
= 4×10^6 Tons CO_{2gas}/day
= 4 million Tons CO_{2gas}/day
= 1,500 million Tons CO_{2gas}/year

One barrel of crude oil produces about 300 kg CO_{2gas} on combustion.
Daily usage of crude oil (average) = 100 10^6 Barrels oil
= 100 $10^6 \times 0.3$ tons CO_2
= 30 million tons CO_2/day

[100 MILLION OIL BARRELS===30 MILLION TONS CO_2]
These estimates indicate that the carbon cycle is significant.

Additionally, it is important to mention that all food/fisheries are a product of carbon dioxide capture (photosynthesis is a typical carbon capture process known on earth). Therefore, food digestion is a carbon recycling process (Appendix).

EXAMPLE:

>PHOTOSYNTHESIS == SUN RADIATION – $CO_{2,gas}$ – WATER<
>FOOD (From land/oceans) PRODUCTION FOR ALL LIVING SPECIES:

CO_{2gas} + water + sunshine ==== photosynthesis === food (land + fisheries
CO_{2gas}>>CO_{2aq} + water + photosynthesis === algae (fisheries)
CO_{2gas}>>CO_{2aq} + === fisheries (fish – shells)

<ALL SHELLS ARE PART OF CARBON CYCLE IN ATMOSPHERE (CO_{2gas}) – OCEAN (CO_{2aq}) CYCLE

In addition, the importance of carbon dioxide, CO_{2gas}, in the atmosphere is evidently very important.

This aspect is briefly mentioned here.

This is generally attributed to the increasing fossil fuel combustion process in technologies. Due to its GHG characteristic, it is thus obvious that mankind should develop and employ technology which can control the increase of $CO_{2,gas}$ from this man-made process .It is also mentioned that since the industrial revolution, mankind has increasingly added $CO_{2,gas}$ to the atmosphere from fossil fuel combustion technology. Furthermore, in combination with increasing use of fossil fuels, industry has also added a higher degree of pollution (e.g.: air-pollution; oceans/lakes/rivers-pollution) (Appendix: A). The combined effect of these processes has been the climate change in different aspects of evolution.

It is also recognized that for immediate mitigation, the carbon-capture-recycling-storage is needed.

The carbon capture recycling and storage (CCRS) technology is currently based on the removal by various processes (Figure 2.1):

> ...**absorption in liquid (fluid);**
> **(by passing-interaction gas through fluid (scrubber))**

> ...**adsorption on solids;**
> (by passing gas through solid particles)

> ...other various techniques:
> (membrane gas separation;

CO_2 –hydrate complex formation; cryogenic method)

In this chapter some of the major processes used for capturing any gas will be described. It is found that the concentration of carbon dioxide as produced (man-made) by fossil fuel combustion is low in different phases:

<Sources of CO_{2gas}:

- atmosphere CO_{2air} (420 ppm: 0.042%)
- flue gases CO_{2flue} (100,000 ppm; 10%)

In order to recycle carbon dioxide (cost benefit) (Birdi, 2020) (Appendix A), one needs to increase the concentration to be almost 100%. The latter object is known to be achieved after using the surface chemistry principles. The purpose of this manuscript is to delineate these different processes (Birdi, 2020).

2.3   CARBON (DIOXIDE) CAPTURE TECHNOLOGIES

Currently, there are various carbon capture – recycling and storage (CCRS) technologies in viable application at various parts of the world (Birdi, 2020; Gates, 2021). The main aim in all these different technologies is to capture any man-made CO_{2gas} from combustion of fossil fuels.

The CO_{2gas} emission control from different industrial (power) plants (flue gases) requires the removal of this GHG gas from the flue gas. In the current literature various studies are reported on the application of absorption/adsorption processes (Myers, 1997; Monson, 2012; Rackley, 2010; Birdi, 2020). A similar approach (with modifications) can be applied to removal/extraction of CO_2 directly from air (Appendix A).

Surface chemistry principles are applied in all of these two (or multi-phase) phase processes (e.g. where a substance A adsorbs/absorbs on another material, B. In the case of gas (Agas) – solid or gas – fluid system:

$$\text{GAS MOLECULE PHASE } (A_{gas})........\text{SOLID/FLUID MOLECULE}$$
$$(B) = \text{ADSORBED/ABSORBED } (A_{gas}) + \text{SOLID/FLUID } (B) \qquad (2.1)$$

In the case of gas – solid system, the adsorbed gas molecules are desorbed by changing the experimental conditions (such as: temperature and pressure). In the case of gas – fluid system, the absorbed gas is desorbed/purged from the fluid by changing the experimental conditions. Both **adsorption** (Section 2.2) and **absorption** (Section 2.3) processes lead to enrichment of the material being captured. For example, the adsorption/absorption process is a method of obtaining enriched gas (>90% CO_2) from a mixture (as from a flue gas (5%–10% CO_2). This may have positive economic consequences in the end use (recycling) of the recovered CO_2.

Pure carbon dioxide (CO_2) is found to exist as gas (CO_{2gas}) at room temperature/1-atmosphere pressure. However, at high pressure and/or temperatures, it turns to liquid ($CO_{2liquid}$) or solid (CO_{2solid}) (Appendix A).

Further, CO_2 in liquid (at high pressure) ca.50 bar)/low temperature = ca. 30°C) (Appendix A)) form is used in various applications in everyday use (such as food industry, cleaning technology, and oil recovery (enhanced: EOR)). This observation is important since the extra expense needed for CCRS can be compensated to some degree by such applications. After carbon adsorption/absorption, it has been recycled or stored (using CCRS technology).

These two different major carbon gas/recycling-capturing methods will be described in this chapter. Furthermore, some additional information about the subject is given (Chapter A; Appendix B).

EXAMPLE

>ABSORPTION OF CO_{2gas} ON SURFACE LAYER OF WATER (E.G. OCEANS/ LAKES/RIVERS)

2.4 ADSORPTION OF GAS ON SOLIDS

In this chapter the processes used for the capturing of carbon dioxide (gas) on solids are described. This is one of the most essential processes recognized to mitigate the increasing emissions.

The interaction of gases with solids has been investigated for over a century (Chattoraj & Birdi, 1984; Adamson & Gast, 1997; Yang, 2003; Bolis et al., 1989; Keller et al., 1992; Birdi, 2017, 2020; Hinkov et al., 2016).

Adsorption of gases on solid surfaces has been investigated, in much detail in the literature. This arises from the fact that this phenomenon is of importance in many everyday technology processes and natural phenomena. One of the most common observations is the corrosion of metals (such as: iron; zinc) when exposed to air. In the corrosion process, oxygen from air interacts with the metal (base metal) to produce metal oxide. Oxygen is also known to interact with different materials on earth and oxidation chemical processes are very important in different cases (both useful and not useful). In fact, corrosion is a very undesirable process, and one of the most costly processes, in every-day life for mankind. Another extensively studied phenomenon is the catalyst technology. In-fact, catalyst technology is also an important surface chemical application technology. On the other hand, one can remove pollutants from flue gases by adsorbing/absorbing the toxic gases such as N_2O_x and SO_2. In other words, in everyday processes, one has these two kinds of gas – solid adsorption processes.

The *gas adsorption on solid*s thus comprises mainly two parts (Figure 2.1):

(INITIAL STAGE)

GAS PHASEGGGG (ADSORBENT)

SOLID PHASESSSSSSSSSSSSSSS (ADSORBATE)

(FINAL STAGE)

...gas molecules (**adsorbent**) ....G G G G G G G G
...solid (**adsorbate**)...................SGSGSGSGSGSGSGS

It is thus seen that the gas molecules are in a different state after adsorption than in the original gas phase. The surface atoms of a solid exhibit due to surface forces to interact with the environment (such as gas molecules (in the present context) and liquids) to varying degrees. **Adsorption** is the process whereby molecules from the gas (or liquid) phase are taken up by a solid surface; it is distinguished from **absorption** which refers to molecules entering into the lattice (bulk) of the solid material or a liquid phase.

The **adsorptive** is the material in the gas phase capable of being adsorbed, whereas the **adsorbate** is the material actually adsorbed by the solid. The solid which exposes the surface sites responsible for the process is called the **adsorbent**. In Figure 2.6 the adsorption process at the surface of a solid material is schematically illustrated (Figure 2.7).

In the case of the air – solid interface, the mass balance can be described as follows. The concentration of the adsorbing gas will thus be expected to be higher near the solid surface, due to adsorption (Figure 2.2). This shows that the adsorption leads to a process where the adsorbed gas is present at a very high concentration, as compared to its original state. This is the most significant effect of the surface adsorption process. The process of all kinds of gas adsorption on solids is governed by either *physical* or *chemical* forces. In the former case, the adsorption is named physical

GAS - SOLID
ADSORPTION PROCESS
MONOLAYER

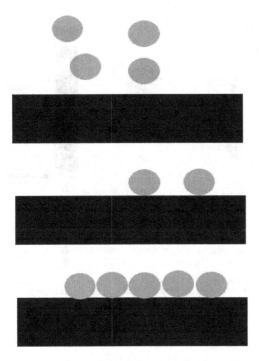

FIGURE 2.6 Adsorption stages of gas on a solid.

adsorption (*physisorption*), whereas in the latter case, it is named chemical adsorption (*chemisorption*) (Appendix A).

The interaction between a gas molecule/atom and a solid is determined by the interaction energy (energies). When two entities (bodies) come in close (i.e. atomic dimension) proximity, there are different kinds of interactions (Figure 2.8). Some of the most common kinds of interaction energies are (Adamson & Gast, 1997; Chattoraj & Birdi, 1984; Bolis, 2001; Birdi, 2003; 2016; 2017; Hinkov et al. 2016) (Figure 2.3):

- dispersion forces (short range);
- electrostatic (long range forces);
- chemical bond.

In the case of *physical* **adsorption** in the present case, the adsorbate (gas) –adsorbent (solid) potential of adsorption will be dependent on various interfacial molecular forces (**dispersion forces** (π_D; **induction forces** (π_I); **electrostatic forces** (π_E)):

$$\pi = \pi_D + \pi_I + \pi_E \tag{2.2}$$

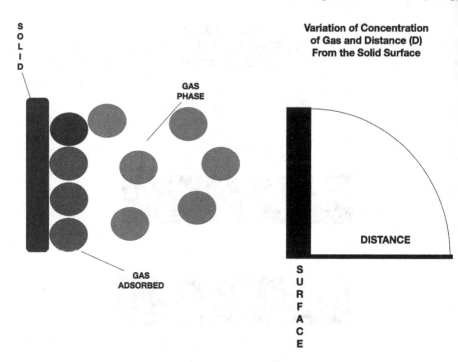

FIGURE 2.7 Variation of density of adsorbed gas on a solid: (a) gas = solid adsorption; (b) gas density.

Where:

    π_D = *dispersion* energy,
    π_I = *induction* energy (interaction between electric field and an induced dipole),
    π_E = *interaction* between electric field (F) and a permanent dipole (μ),

(Adamson & Gast, 1997; Chattoraj & Birdi, 1984; Birdi, 2003, 2016, 2020),
which are operative in all adsorbate – adsorbent systems. The quantity π_E potential interaction contribution arises from charges (on the gas and solid surface) (Birdi, 2020). However, in the case of chemical interactions (where a chemical bond is formed), the process is very different. For instance, the oxidation of metals (such as iron; aluminum) by oxygen (as found in air) leads to the formation of a new compound (e.g. Fe_2O_3; Al_2O_3). Of course, when the concentration of gas is very low (as in the case of CO_{2gas} in air), chemical bond formation is preferable to other techniques.

The interaction potentials are related between an atom/molecule (or a charge) on the surface and the adsorbate atom/molecule. All interactions between two bodies depend primarily on the magnitude of the distances separating these, R_d.

The dispersion forces are given as:

$$\pi_D = -A_D R_d^{-6} + B_D R_d^{-12} \qquad (2.3)$$

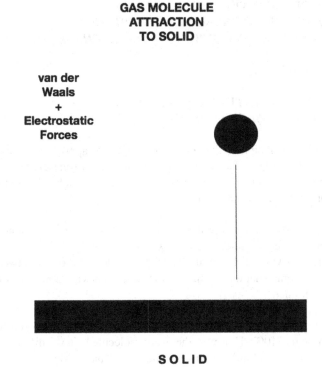

**GAS MOLECULE
ATTRACTION
TO SOLID**

**van der
Waals
+
Electrostatic
Forces**

SOLID

FIGURE 2.8   Gas – solid interactions (different surface forces).

where A_D and B_D are constants. The expression for π_I (field (of an ion) and induced point dipole):

$$\pi_I = -2\ \alpha\ F \qquad (2.4)$$

Gas adsorption on a solid is an analogous process of separation of molecules (such as gas chromatography methods) (Birdi, 2020). The energetics of the process is thus dependent on the gas phase and solid phase characteristics. The degree of interaction between the molecules in the gas phase and solid thus can be investigated. As regards the distance between molecules/atoms in different phases can be estimated as follows. The different phases of matter can be described as follows.

In the case of all natural processes, one defines the different phases of matter:

- **Gas**
- **Liquid**
- **Solid**

Further, when any two or more phases meet, one has a **surface** (or **interface**) to consider (Chapter 3; Appendix A).

SOLID PHASE//NTERFACE/GAS PHASE
SOLID PHASE//NTERFACE/LIQUID PHASE
SOLID(1) PHASE//NTERPHACE/SOLID(2) PHASE

And in the case of liquids:

LIQUID PHASE/INTERFACE/GAS PHASE
LIQUID (1) PHASE/INTERFACE/LIQUID(2) PHASE

Surface chemistry principles (Chapter 3 & Appendix A) apply to the state of properties of these different "interfaces". The state of the matter (phase) at any given temperature or pressure is given by the distance between the molecules and the energy of interactions between the molecules.

> SOLID PHASE: The molecules in the solid state are closely bound to each other and molecular forces keep the shape of a solid.
> LIQUID PHASE: In the liquid state, the molecular forces between molecules are somewhat weaker than in the solids. Thus a liquid takes the shape of its container.

In general, the distance between molecules in the solid phase is shorter than that in the liquid phase (ca. 10%). However, the water molecule behaves differently. The ice is lighter (ca. 10%) than liquid water (Appendix A). That is why ice always floats on water (for example: icebergs float on water).
In order to explain this in more semi-quantitative detail, let us consider water.
Water is found in the liquid state at room temperature and pressure (1 atmosphere).

EXAMPLE

>Simple estimate of the molecular state of solid (ice) and liquid (water), in terms of packing (i.e. distance between molecules).
In order to estimate the distance between the molecules the following data are useful (standard state):

1 mole of water (18 g) (H_2O)(at room temperature and pressures)
Water density $= 1.0$ g/cc
Volume of one mole (water) $= 18$ cc/mol
Volume of 1 mole of gas (all gases) $= 22.4 \times 1,000$ cc $= 22.4$ L $= 22,400$ cc
Ratio of volume of gas: liquid $= 22,400/18 =$ ca. 1,000 times

(Note: This is a very simple but useful estimate as regards the molecule-molecule distance estimates in different states of matter.)
    (In general: volume density per molecule/atom in gas is ca. 1,000 x larger.)

This simple analysis shows that in general, a molecule in the gas phase occupies **1,000** times more volume than that in the liquid/solid phase (see Chapter 1).

This means that the molecules in the gas phase move about larger distances (ca. 10 times) than in liquid or solid phase before collision. Further, in general, in the solid phase the molecule occupies ca. 10% less volume than in the liquid phase. However, water is an exception, which shows that ice is ca. 10% lighter than water at zero degrees (hence ice (iceberg) floats on water). The latter observation has other consequences, as regards absorption of gases in ice (Appendix A). Consequently, this also makes the oceans/lakes additional physical properties as regards absorption and diffusion of gases (especially CO_{2gas} which is slightly soluble in water) (Appendix A).

This shows that any process, which converts the system from a gas phase to a liquid (solid phase, decreases the volume per mole to about by a factor 1,000.

In general, any **adsorption** is a process where one substance (atoms or molecules) of one substance in one phase (in the present case gas/ fluid phase) interacts (physical

---

EXAMPLE

<GAS – SOLID adsorption >
<GAS PHASE – > GAS-SOLID PHASE

---

interaction or chemical interaction) with the surface of a different phase (the surface of a solid).

2.5   THEORY OF ADSORPTION OF GAS ON SOLID SURFACES (BASIC REMARKS)

Currently, one finds that the *equilibrium concentration of carbon* (CO_{2gas}) in atmosphere (air) is very low (currently 400 ppm – 0.04%). Some studies have shown that this magnitude is increasing since the industrial revolution (Birdi, 2020), mainly due to increasing energy demand from population and therefore increasing usage of fossil fuels (combustion) (Birdi, 2020).

In any gas phase, the molecules are known to be in continuous movement and possess kinetic energy (kB T). Furthermore, the gas molecules experience surface forces when these come in proximity of a solid (surface). The experiment shows that that the gas molecule (or atom) may adsorb on the solid surface with varying surface forces. The degree of gas adsorption is found to be determined by solid surface forces (Chapter 3).

It was suggested that as regards the continuous change from the liquid to the vapor state, at temperatures above the critical (based on van der Waals theory: at the boundary between a liquid and its vapor there is not an abrupt change from one state to the other, but rather that a transition layer exists in which the density and other properties vary gradually from those of the liquid to those of the vapor (Langmuir, 1917; Adamson & Gast, 1997; Meyers & Monson, 2000; Chattoraj & Birdi, 1984;

Keller & Staudt, 2006; Birdi, 2003, 2020). Later, this postulate of the continuous transition between phases of matter has been applied generally in the development of theories of surface phenomena, such as surface tension, adsorption, absorption, catalysis, foams, and bubbles (Adamson & Gast, 1997; Keller et al., 1992; Birdi, 2003, 2020). A simple theory has been used to describe the surface tension phenomena, as follows. . One assumes that the molecules in the transition layer are attracted towards one another with a force which is an inverse exponential function of the distance between them. Accordingly, experiments show that there is an abrupt change in physical properties in passing through the surface of any solid or liquid. The atoms/ molecules which are in the surface of a solid are held to the underlying atoms by forces similar to those acting between the atoms inside the bulk of the solid phase. From crystal structure studies and from many other considerations, it is known that these forces are of the type that have usually been designated as chemical. In the surface layer, because of the asymmetry of the conditions, the arrangement of the atoms must always be slightly different from that in the interior of the bulk phase. It is seen that these atoms will be unsaturated chemically and thus they are surrounded by an intense field of force.

The adsorption process is described as:

<that when gas molecules impinge against any solid or liquid surface they do not in general rebound elastically, but may adsorb on the surface, being under the field of force of the surface atoms. Additionally, gas molecules loose kinetic energy after desorption. These gas (adsorbed) molecules may subsequently desorb from the surface of the solid (Figure 2.9):

This is indirectly observed from the fact that the degree of adsorption is found to increase at lower temperatures (i.e. decrease in kinetic energy).

- GAS MOLECULES → MOVE (IN THREE DIMENSIONS) WITH KINETIC ENERGY
- GAS MOLECULE --> IMPINGES ON SOLID
- GAS MOLECULE BOUNCES BACK <—> NO ADSORPTION
- GAS MOLECULE ADSORBS ON SOLID SURFACE ⇔ ADSORPTION

The gas molecule/atom moves with kinetic energy towards the solid. In the case of absence of adsorption, it bounces back (Figure 2.9). However, if the adsorption energy is higher than the kinetic energy, then adsorption is observed (Figure 2.9).

The length of time that elapses between the adsorption of a molecule and its subsequent desorption will depend on the degree of intermolecular surface forces (Figure 2.4). Further, in case that the surface forces are relatively strong, desorption will be observed at a very low rate. This will lead to the complete surface coverage with a layer of molecules. In cases of gas adsorption, this layer will usually be not more than one molecule deep, for as soon as the surface becomes covered by a single layer (monolayer) the surface forces are chemically saturated. On the other hand, if the surface forces are weak the desorption may take place. This means after the adsorption, only a small fraction of the surface becomes covered by a single layer of adsorbed gas molecules. In accordance with the chemical nature of the surface forces, the range of these forces has been found to be extremely small, of

Gas-Solid
Adsorption

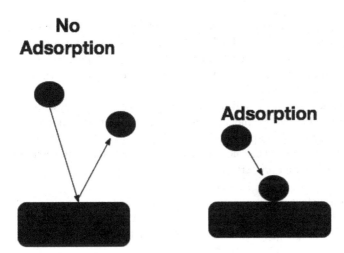

No Adsorption

Adsorption

FIGURE 2.9   Adsorption characteristics (positive/negative) of a molecule of gas on a solid.

the molecular order of 10^{-8} cm. That is, the effective range of the surface forces is usually much less than the diameter of the gas molecules. The gas molecules thus usually orient themselves in definite ways in the surface layer since they are bound to the surface by forces acting between the surface and particular atoms or groups of atoms in the adsorbed molecule. The orientation of adsorbed molecules on the solid surface is thus another specific characteristic (as found from atomic microscope studies) (Birdi, 2020).

The dipole of gas molecules will also determine the absorption interaction energy.

It is estimated that the average diameter of molecules of ordinary permanent gases averages about 3×10^{-8} cm (3 Å). There are thus about 10^{15} molecules per cm^2 in a monomolecular layer, corresponding to 0.04 cubic millimeter of gas per cm^2, at ordinary temperature and pressure; therefore, with surfaces of about a square meter, the amount of gas required to cover the surface with a single layer of molecules is 400 cubic millimeters (Figure 2.1).

This model (ideal) for gas adsorption on solids is shown as follows:

GGGGGGGGGGGGGGGGGGGGGGGGG.adsorbed gas molecules
SSSSSSSSSSSSSSSSSSSSSSSSSSSSSS.surface molecules of solid
SSSSSSSSSSSSSSSSSSSSSSSSSSSSSS.bulk solid molecules
SSSSSSSSSSSSSSSSSSSSSSSSSSSSSS

It is thus seen that

- distance between G molecules (i.e. between **G – G**) in the adsorbed state is almost ten times less than that in the gas state.
- distance between G and S (i.e. **G – S**) is also reduced after adsorption (fewer degrees of motion).

Extensive gas adsorption studies on different solid surfaces (e.g. mica, glass, and platinum; zeolites; carbon; and coal) have been carried out (Chapter 2).

For example, in the case of solids, such as platinum, it is known to form a non-oxidizing metal surface. This provides useful adsorption data.

The adsorption potential will be dependent on the surface forces at the interface: gas molecule/atom – solid.

The interaction energy will thus be different for different systems. Therefore, in any industrial application, one needs to know these interaction potential to be able to achieve the maximum output.

The surface of a solid can be characterized (as regards the molecular topography) by following different properties:

- **plane** surface,
- **rough** surface,
- **porous** solids (pores with varying sizes/shapes) (Figure 2.10).

<GAS – SOLID ADSORPTION ON POROUS SOLIDS>

The **plane** faces of a crystal are found to consist of atoms forming a regular plane lattice structure. The atoms in the cleavage surface of crystals like mica are those which have the weakest fields of surface forces of any of the atoms in the crystal. It is probable that in mica the hydrogen atoms cover most, if not all, of the surface, since hydrogen atoms, when chemically saturated by such elements as oxygen, possess only weak residual valence. In the case of solid surfaces such as glass, and other oxygen compounds (e.g. quartz; calcite; zeolites), the surface probably consists of a lattice of oxygen atoms. The surface of crystals thus resembles, to some extent, a well-defined surface for gas adsorption sites. When molecules of gas are adsorbed by such a surface these molecules take up definite positions with respect to the surface lattice and thus tend to form a new lattice above the old. The new structures of such an adsorbed phase can be studied by suitable spectroscopic analyses (Chapter 4).

It is found that an unit area of any crystal surface, therefore, has a definite number of gas adsorption sites with each one selectively able to adsorb the gas molecule or atom. In general, these specific binding sites will not all be expected to be alike, i.e. as regards the adsorption energy. It is found that there will frequently be cases where there are two or three different kinds of sites. For example, in a mica crystal it may be that both oxygen (O) and hydrogen (H) atoms, arranged in a regular lattice, form the surface in such a way that different elementary spaces are surrounded by different numbers of arrangement of atoms.

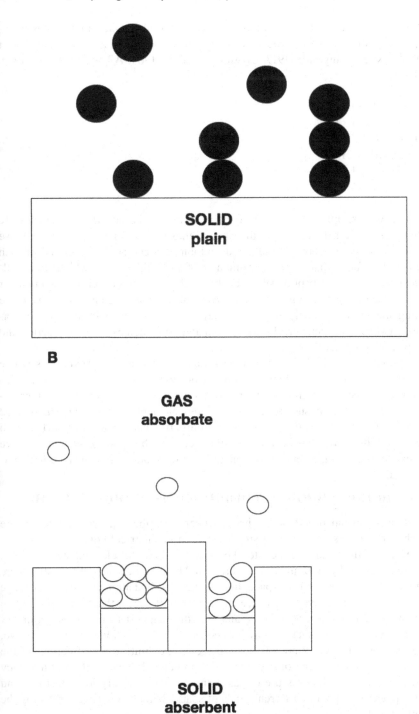

FIGURE 2.10 Gas adsorption on smooth or porous solid surface.

It is useful to consider an example: one may consider a (schematic) most plausible arrangement of oxygen (O) and hydrogen (H) atoms in a surface lattice (for example mica) (Langmuir, 1917; Chattoraj & Birdi, 1984; Adamson & Gast 1997; Birdi, 2020):

H..O..H..H..O..H
H..H..O..H..H..O
0..H..H..0..H..H
H..0..H..H..0..H
H..H..O..H..H..O
O..H..H..O..H..H

This is a very simplified example of real systems. If the adsorbed molecules are present at sites over the centers of the squared thus tend to form a new lattice above the old. Experiments show that all solids exhibit very characteristic gas adsorption behavior. It is found that there is per unit area of any solid surface, which has a well-defined number of "adsorption sites". Each adsorbing site has a well-defined number of gas molecules/atoms. In general, it is reasonable to expect that not all of the active binding sites will be exactly alike. This may arise from defects in the surface morphology (which has been corroborated) from analyses (such as X-ray diffraction and scanning atomic microscopy).

It is known that there are also some solids with surface characteristics where there are two or three different kinds of adsorption sites. For example, in a mica crystal it may be that both oxygen and hydrogen atoms, arranged in a regular lattice, form the surface in such a way that different elementary spaces are surrounded by different numbers or arrangements of atoms. In-fact, this kind of surface will be expected in all solid minerals. For example, it has been suggested that the one may expect the arrangement of oxygen and hydrogen atoms in a surface lattice as indicated:

HOHHOH HHOHHO 0HH0HH H0HH0H HHOHHO OHHOHH

The state of the adsorbed molecules has been described. In the systems where adsorbed molecules are found at sites (based on geometrical considerations) over the centers of the square shape site, there are two kinds of elementary situations, those represented by **H H** and those represented by **0 H.** For each of the latter there are two of the former kinds of space. In case the adsorbed atoms/molecules arrange themselves directly above the surface atoms, one will expect that there may be two kinds of gas adsorption sites. These analyses thus suggest from considerations of this kind that a crystal surface may have sites of only one kind, or may have two, three, or more different kinds of sites representing definite simple fractions of the surface. Further, each kind of site space will, in general, have a different tendency to adsorb gases. As the pressure of gas is increased the adsorption will then tend to take place in steps, the different kinds of spaces being successively filled by the adsorbed molecules.

One may thus safely conclude that the adsorption of a gas on a solid may show many different types of adsorption phenomena. This is indeed also observed from experimental data (Chapter 4).

2.6   DIFFERENT METHODS USED FOR GAS ADSORPTION STUDIES

In general, in a system where a solid is exposed in a closed space to a gas at pressure p, the weight of the solid typically increases (due to gas adsorption: as shown in Figure 2.6) and the pressure of the gas decreases: the gas is adsorbed by the solid. At equilibrium, the magnitude of pressure peq does not change and the weight reaches an equilibrium value. A schematic of a simple adsorption setup is given in Figure 2.11. Many commercial apparatus are available which are used for measuring the gas adsorption (at varying temperature and pressure) (Chapter 2).

The quantitative amount of gas adsorbed on a solid is experimentally determined:

   i. by gravimetry (for example: the increase in weight of the solid is monitored by any suitable and sensitive balance);
   ii. by volumetry (the fall in the gas pressure is monitored by manometers/transducer gauges);
   iii. by monitoring the change of any other physical parameter related to the adsorption of matter, such as the evolved heat (if the heat of adsorption is known and constant) or the integrated IR absorbance (if the specific molar absorbance of adsorbed species is known).

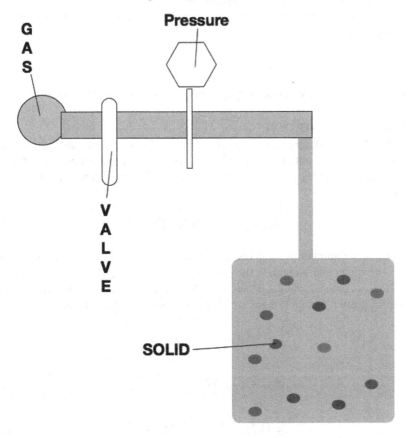

FIGURE 2.11   Schematics of typical apparatus used for gas adsorption on a solid.

A simple adsorption process of gas molecules on a plane surface having only one kind of elementary site and in which each site can only adsorb one gas molecule/atom can be described as follows.

2.7 ANALYSES OF GAS ADSORPTION ON SOLIDS (KINETIC MODEL)

There is also one of the gas adsorption process which has been defined by the kinetic theory.

The gas – solid adsorption phenomenon is a dynamic molecular system.

Molecules in the gas phase are in continuous motion (as follows from the kinetic theory of gases (Adamson & Gast, 1997; Birdi, 2003, 2020). It is thus obvious that all processes involving gases would be dynamic phenomena. Gas molecules will thus continuously strike the surface of a solid. Depending on the experimental conditions (pressure; temperature), there may happen any of the following (Figure 2.2.):

- gas molecules bounce back to the gas phase (weak or no adsorption)
- gas molecule may adsorb.
- gas molecules may adsorb = desorb continuously.
- gas molecules may interact (e.g. covalent bonding; steric bonding) with the solid surface.

DYNAMICS OF GAS – SOLID SURFACE:

GAS MOLECULES CONTINUOUSLY STRIKE SOLID SURFACE:

DEGREE OF ADSORPTION
From this dynamic description of the gas adsorption process, one can derive the relation between the amount of gas adsorbed and the pressure and temperature. The rate at which gas molecules come into contact with a surface is given by the relation for kinetic theory:

$$N_{gas} = (M_{gas} / (2\pi R T)^{0.5} p_{gas} \qquad (2.5)$$

Here N_{gas} is the number of grams of gas striking the surface per sq. cm. per second, M_{gas} is the molecular weight, T the absolute temperature, p_{gas} the pressure in bars, and R the gas constant 83.2×10^6 ergs per degree. If one defines a quantity g_{gm}, the number of gram molecules of gas striking each cm^2 per second, then:

$$g_{gm} = N_{gas} / M_{gas} \qquad (2.6)$$

$$g_{gm} = p_{gas} / \left(2 \pi M_{gas} R T \right)^{0.5} \qquad (2.6a)$$

$$= 43.75 \ 10^{-6} \ p_{gas} / \left(M_{gas} T \right)^{0.5} \qquad (2.7)$$

This is a simple modification of the relation giving the rate of effusion of gases passing through small openings (Adamson & Gast, 1997; Birdi, 2003, 2020).

This system may be described as:

<INITIAL STATE:

...gas phase (gas molecules G)

...solid phase (solid molecules S)

AFTER GAS ADSORPTION:

...GAS PHASE (GAS MOLECULES)

...GAS ADSORBED

ON SOLID...GGGGGGGGGGGGGGGG

SSSSSSSSSSSSSSSSSSSSS

The state of molecules in the gas phase has been analyzed by the classical kinetic gas theory (Adamson & Gast, 1997; Chattoraj & Birdi, 1984; Keller & Staudt, 1992; Yang, 2003; Bolis et al., 1989; Myers & Monson, 2000).

Gas molecules move about in space and possess kinetic energy and occupy 1000 times more volume than in liquid or solid state (equation 2.2). It is thus obvious that the state of adsorption on a solid is where the gas molecules are present comparatively with very low kinetic energy.

It is also assumed that the gas molecules only adsorb on specific sites (N_S) on the solid (Chapter 3). Furthermore, all solids have a definite maximum number of sites on the surface (per gram of solid), where gas molecules can adsorb. This number can be estimated by an adsorption method (Bolis et al., 1989; Adamson & Gast, 1997).

The fractional monolayer coverage (θ) of the sites occupied by adsorbate molecules is defined as:

$$\theta = N_s / N_{total} \qquad (2.8)$$

where N_{total} is total number of absorbing sites on the solid. The quantity N_s is a unique characteristic of solid. The rate of adsorption is given as:

$$\text{adsorption rate} = k_{ads}\, p_{gas}\, (1-\theta) \qquad (2.9)$$

kads being the rate constant for the adsorption and $(1-\theta)$ the fractional monolayer coverage of sites not occupied yet by the adsorbate molecules.

The *rate of gas desorption*, k_{des} as related to the rate constant for desorption, is given by the equilibrium (dynamic conditions):

$$\left(\text{adsorption rate}\right) = \left(\text{desorption rate}\right) \qquad (2.10)$$

From this one gets (the Langmuir gas-adsorption equation):

$$\text{desorption rate} = k_{des}\theta \qquad (2.11)$$

$$k_{ads}\, p_{gas}\, (1-\theta) = k_{des}\, \theta \qquad (2.12)$$

$$\theta/(1-\theta) = K\, p_{gas} \qquad (2.13)$$

where:

K = ratio of the rate constant for adsorption/ the rate constant for desorption,
= k_{ads} / k_{des}

The gas adsorption data are analyzed (*Langmuir equation*: Chattoraj & Birdi, 1984; Birdi, 2020) by the relation:

$$\theta = V/V_{mon} = \left(K\, p_{gas}\right)/\left(1 + K p_{gas}\right) \qquad (2.14)$$

The term V represents the adsorbate volume and V_{mon} the monolayer volume, i.e. the volume of adsorbate required to complete the monolayer.

At very low pressure, the equation reduces to a linear dependence of the coverage upon the equilibrium pressure ($\theta = hp$). Conversely, at high pressure the equation reduces to the case of coverage approaching the monolayer ($\theta \approx 1$).

The quantity monolayer coverage (Vmon), the quantitative magnitude of this is not easily determined experimentally with high accuracy. Hence, by using a different procedure: i.e. the Langmuir equation is suitably transformed in the so-called reciprocal linear form:

$$(1/V) = 1/(K\, V_{mon})(1/p_{gas}) + 1/V_{mon} \qquad (2.15)$$

In the case where the data fit the Langmuir model, the reciprocal volume a_1/V against reciprocal pressure 1/p plot is linear (Langmuir-type isotherms). In other words, such data plots provide useful information as regards the gas adsorption equilibrium.

On the other hand, if the experimental data plot deviates from this linear relation, it indicates that the Langmuir equation does not agree with the given adsorption process. This observation is analogous to the ideal gas equation. The deviation of the adsorption isotherm thus indicates a non-ideal system.

The magnitude of the quantity monolayer capacity is estimated from the intercept $i = 1/V_{mon}$ of the straight line. From this one can estimate the equilibrium constant K from the value of the quantity slope (s_{slope}):

$$s_{slope} = 1/(K\, V_{mon}) \qquad (2.16)$$

The monolayer volume and the equilibrium constant are typical of the adsorbent/ adsorbate pairs at a given temperature. In particular, the value of K is related to the strength of the adsorbent-adsorbate interaction: high values of K indicate large strength, and low values weak adsorbing forces.

As an example: the data of a typical system, i.e. case of CO adsorbed at $T = 303$ K on dehydrated Na— and K–MFI will be discussed (as regards the experimental volumetric and calorimetric isotherms). The number of CO molecules adsorbed per gram of zeolite at p_{CO} represents the number of occupied sites (NS), whereas the number of charge-balancing cations exposed per gram of zeolite represents the total available sites (N).

From a data plot of the coverage $\theta = N_S/N_{total}$ and the $\theta/(1-\theta)$ quantity are plotted against p_{CO}. The slope of the $\theta/(1-\theta)$ versus p_{CO} plot is the Langmuir constant (Figure 2.12) K). The CO_{gas} adsorbing molecules are found to exhibit soft <Lewis> base properties (i.e. it gets polarized by the electrostatic field generated by the alkaline-metal cations located in the MFI zeolite nano-cavities. This leads to the adsorption of gas molecules (reversibly) on the surface when in contact with the (activated) zeolite [23].

The equilibrium constant K_{ads} for solid Na–MFI (4.88 ± 0.02 10^{-3} Torr^{-1}) was found to be larger than that for K–MFI (1.15 ± 0.02 10^{-3} Torr^{-1}). This was found to be in agreement with the different polarizing characteristics of the cations (e.g. Na$^+$ & K$^+$). In fact, the local electric field generated by the coordinatively unsaturated cations depends on the charge/ionic radius ratio, which is larger for Na$^+$ than for K$^+$, (ionic radius of Na$^+$ = being 0.97 Å and of K $^+$ = 1.33 Å). As regards the charge/ionic radius ratio, the maximum coverage attained at $p_{CO} = 90$ Torr was larger for Na–MFI ($\theta \approx 0.3$) than for K–MFI ($\theta \approx 0.1$). This shows that there is a correlation between adsorption and the size of the cation.

For example: The analyses of the adsorption data for CO further showed: The magnitude of (standard free energy) ΔG°_{ads} for CO adsorption at the two alkaline-metal sites was obtained from the Langmuir equilibrium constant K:

$$\Delta G^{\circ}_{ads} = -R\,T\left(\ln K_{ads}\right) \qquad (2.17)$$

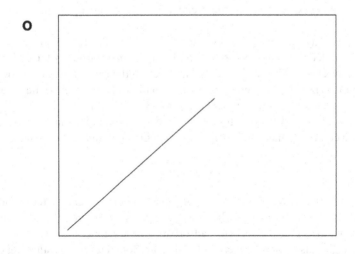

O

O/(1 - O)

FIGURE 2.12 θ versus $\theta/(1-\theta)$.

In both cases the adsorption process in standard conditions was found to be *endothermic*:

$$\Delta G_{ads}^{\circ} = +13.4 \text{ kJ mol}^{-1} \text{ for Na} - \text{MFI};$$

and

$$\Delta G_{ads}^{\circ} +17.0 \text{ kJ mol}^{-1} \text{ for K} - \text{MFI}.$$

Further studies also showed that after vacuum treatment the two: $Na^{+} \cdots CO$ and $K^{+} \cdots CO$ adspecies were absent. This shows that CO was absent after the vacuum treatment of the solid.

From ΔG_{ads}° the standard entropy of adsorption ΔS_{ads}° can be estimated, if ΔH_{ads}° is known. The CO adsorption enthalpy change was measured calorimetrically during the same experiments in which the adsorbed amounts were measured.

Generally, the gas adsorption data for real systems show deviation from the ideal Langmuir adsorption model (especially at high gas pressures) (Chapter 4): According to the assumptions of Langmuir theory:

i. the solid surface is rarely uniform: there are always "imperfections" at the surface,
ii. the mechanism of adsorption is not the same for the first layer of molecules as for the last to adsorb. When two or more kinds of sites characterized by different adsorption energies are present at the surface (as stated in point i), and when there are lateral interactions among adsorbed species occur (as stated in point ii), the equivalence/independence of adsorption sites assumption fails. The most energetic sites are expected to be occupied first, and the adsorption enthalpy ΔH (per site) instead of maintaining a constant, coverage-independent value, exhibits a declining trend as far as the coverage θ increases.

Further, one may also find that in some systems: on the top of the monolayer of gas, other molecules may adsorb and multi-layers may build up the (BET) model (Adamson & Gast, 1997; Birdi, 2003). However, in the case of the adsorption at surfaces characterized by a heterogeneous distribution of active sites, the results were found to be different (Adamson & Gast, 1997; Birdi, 2003).

Furthermore, one has also found that the gas adsorption data on a solid (the Freundlich equation) may exhibit an isotherm which will fit the following expression:

$$V_{ads} = kp_{gas} \, 1/n \tag{2.18}$$

This relation is based on simple empirical considerations, where the term Vads represents the adsorbed amount, pgas the adsorptive pressure, whereas k and n are empirical constants for a given adsorbent - adsorbate pair at temperature T.

In some gas adsorption isotherms, the data are found to fit the following equation (an exponential equation: Temkin isotherm):

$$V_{ads} = k_1 \ln \left(k_2 \, p_{gas} \right) \tag{2.19}$$

It is known that this relation is a purely empirical formula, where V_{ads} represents the adsorbed amount and p the adsorptive pressure; k_1 and k_2 are suitable empirical constants for a given adsorbent-adsorbate pair (at temperature T). The isotherm data suggest that the adsorption enthalpy H_{ads} Δ (per site) decreases linearly upon increasing adsorption.

Examples of heats of adsorption decreasing linearly with coverage are reported in the literature, as for instance in the case of NH3 adsorbed on hydroxylated silica, either crystalline, or amorphous. Similar results were obtained in the case of CH3OH adsorption on silica-based materials.

Further, it is found that at sufficiently low pressure all gas adsorption isotherms are linear and may be regarded as obeying the <Henrys> law (Figure 2.7):

$$V_{ads} = h_H \, p_{gas} \qquad (2.20)$$

In this model, the <Henrys> law of gas adsorption is considered to relate to the amount of gas adsorbed to the pressure, p_{gas}. The Henry constant, h_H, is typical of the individual adsorbate-adsorbent pair, and is obtained from the slope of the straight line representing the isotherm at low adsorption coverage. The isotherm classification, which is of high merit in terms of generality, deals with ideal cases which in practical work are rarely encountered. In fact, most often the adsorption process over the whole interval of pressure is described by an experimental isotherm which does not fit into the classification.

Regardless, each of the equations described above may be applied over a given range of equilibrium pressure. This procedure thus allows one to analyze the experimental isotherm through the combination of individual components to the process. By using this approach, the surface properties of the solid, and the thermodynamics features of processes taking place at the interface can be quantitatively analyzed (Adamson & Gast, 1997; Birdi, 2020).

As an example: the adsorption data of NH_3 on a dehydrated silica specimen was found to be consistent with the Langmuir and Henry isotherms equations. This behavior was explained due to the presence of H−bonding on silanol (Si-OH) groups.

<NUMBER OF ADSORBING SITES (FOR GAS MOLECULES) ON A SOLID:
It is also important to determine as regards the available number of sites on a unit surface area of a solid.

Experiments have shown that this is one of the most characteristic of any solid surface,

All solid surfaces are found to exhibit a specific **maximum** number of adsorption sites (depending on the gas and the solid). This quantity can vary from zero to a specific number of adsorption sites.

This is also found from experiments. One may denote N_o as the number of elementary sites per cm^2 of surface. Hence, the number of gas molecules adsorbed cannot exceed N_o, except by the formation of additional layers (multi-layers) of molecules. The forces acting between two layers of gas molecules will usually (expected) be very much less than those between the solid surface and the first layer of molecules. In any case, the two cases being different, e.g. G – S or G – G, will be expected to be under different interactions. This means that the energy of interaction is different:

GGGGGGGGGGGGGGG
SSSSSSSSSSSSSSSSSS
SSSSSSSSSSSSSSSSSSS
SSSSSSSSSSSSSSSSSSS

The structure of *double-layer adsorption*:

GGGGGGGGGGGGGGG
GGGGGGGGGGGGGGG
SSSSSSSSSSSSSSSSSSSS
SSSSSSSSSSSSSSSSSSSS

<MOLECULAR THEORY OF GAS ADSORPTION ON SOLIDS:
It is useful to consider the molecular basis of adsorption in the CCRS technology.

In any adsorption phenomenon the substances from the external environment (e.g. gas or liquid) are absorbed by a solid surface (adsorbent). The adsorption process has been used to separate gaseous and liquid mixtures, for drying and purification of gases and liquids. The adsorption calculation of the equilibrium and dynamic characteristics of adsorption in porous bodies at the molecular level have been investigated in the literature.

In a recent study adsorption theories based on statistical physics were derived to explain the adsorption process (Tovbin, 2017).

These were found to be the consistent with the description of the equilibrium distribution of molecules and dynamics of flows in complex porous materials. These data were found to be useful for a wide range of practical applications in the development of new technologies.

Furthermore, the effect of gas pressure on adsorption is found to be as follows:

> at low gas pressure: the amount of adsorbed gas is proportional to the pressure, but increases much more slowly at higher pressures. However, if the relative adsorption rates for the two gas species are different from each other, then, experiments show that the adsorption isotherms show definite steps as the pressure increases.

2.8   SOLIDS WITH SPECIAL CHARACTERISTICS FOR GAS ADSORPTION

In everyday one finds solids which one uses exhibit with different structures. Some of the different solid characteristics are given in the following.

> **Adsorption on amorphous surfaces**: In general with crystal surfaces there are probably only a few different kinds of elementary spaces, but with amorphous substances (such as glass; zeolites), the elementary binding sites may

all be different. One may thus expect the surface divided into infinitesimal fractions, with each possessing its own binding site.

Furthermore, one finds gas – solid systems with some very unique properties. Experiments show that the surface forces which hold adsorbed substances act primarily on the individual atoms rather than on the molecules. When these forces are sufficiently strong, it may happen that the atoms leaving the surface become paired in a different manner from that in the original state molecules. These examples of gas adsorption are found in different *catalysts*. The synthesis of ammonia (NH_3) from nitrogen and hydrogen (by using a catalyst) is one of the most important examples.

However, in the case of the adsorption of a diatomic gas such as oxygen (O2: $O = O$) or carbon monoxide (CO: $C = O$) on a solid surface different analyses are needed. If one assumes that the atoms are individually held to the solid surface, each atom occupies one elementary space. The rate of desorption of the molecules is expected to be negligibly small, but occasionally adjacent atoms combine together and thus nearly saturate each other chemically, so that their rate of evaporation becomes much greater. The gas molecules thus leave the surface only in pairs. For example, initially the system consists of a solid surface with no adsorbed gas molecules.

Starting with a bare solid surface, if a small amount of gas is adsorbed, the adjacent atoms will nearly always be the atoms which exhibit adsorption together when a molecule was adsorbed. In some cases, however, two molecules will happen to be adsorbed in adjacent spaces. One atom of one molecule and one of the other may then desorb from the surface as a new molecule, leaving two isolated atoms which cannot combine together as a molecule and are therefore compelled to remain on the surface. At equilibrium, one may expect that there will be a haphazard distribution of atoms over the surface.

Furthermore, there may also be systems of gas - solid surface, where there will be two kinds of adsorbing sites:
one of which are occupied by the gas molecules:

GGGGGGGGGGGGGGGG..................gas phase
SSSSSSSSSSSSSSSSSSS.................solid surface phase
SSSSSSSSSSSSSSSSSSS
SSSSSSSSSSSSSSSSSSS

while there will be un-occupied sites (indicated as ...):

GGG....GGG....GGG....GGG
SSSSSSSSSSSSSSSSSSSSS
SSSSSSSSSSSSSSSSSSSSS
SSSSSSSSSSSSSSSSSSSSS   .

In order that a given molecule approaching the surface may adsorb (and be retained for an appreciable time) on the surface, two particular elementary sites must be vacant. The gas adsorption thus becomes dependent on the chance of the molecules hitting the two kinds of site.

Adsorption of more than one gas molecule in thickness **(multi-layer adsorption) on solids:** With gases or vapors at pressures much below saturation the surface of a solid tends to become covered with a single layer of molecules. The reason for this is that the forces holding gas molecules (or atoms) on to the surface of solids are generally much stronger than those acting between one layer of gas molecules and the next. When the vapor becomes nearly saturated, however, the rate of evaporation from the second layer of molecules is comparable with the rate of condensation so that the thickness of the gas **film** may exceed that of a molecule. These thicker films may also be present in those cases where the forces acting between the first and second layers of adsorbed molecules are greater than those holding the first layer to the surface. An example of this latter kind has been found experimentally in the condensation of cadmium vapor on glass surfaces.

2.9   GAS – SOLID ADSORPTION ISOTHERMS AND MECHANISMS

Currently, the CCRS technology is being developed to mitigate the different effects of fossil fuel usage: worldwide. Hence, there is a need to investigate the mechanisms involved in the gas – solid adsorption process mechanism. Experiments have shown that the adsorption process at the gas/solid interface is related to an enrichment of one or more components in an interfacial layer (Chattoraj & Birdi, 1984; Adamson & Gast, 1997; Keller & Staudt, 2006; Birdi, 2003, 2020):

GAS MOLECULE (G). G.G.G.
ADSORBED GAS MOLECULE (G)
GGGGGGGGGGGGGGGGGGGGG
SOLID SURFACE SOLID SURFACE

The mechanism of gas adsorption (e.g. amount of gas adsorbed per unit gram solid (with a specific area/gm); number of layers of gas adsorbed) is determined from the analyses of experimental data isotherms. The gas adsorption will be dependent on temperature and pressure, for each system. The adsorption mechanisms, e.g. monolayer; bi-layer; multi-layer, will lead to different kinds of isotherms (Figures 2.13–2.16). This arises from the fact that the energetics of adsorption is dependent on the arrangement of gas molecules. These literature data have shown that (Sing et al., 1985), there exist distinct **six different types of adsorption** systems (Figure 2.7a and b).

The mechanism of gas adsorption (e.g. monolayer or bi-layer or multilayer) on any solid surface can be determined from the experimental data. The **shape of the adsorption isotherm** has been found to provide useful information as regards the physical characteristic of the adsorbate (gas) and solid adsorbent (in case of porous solids the pore structures) geometry (Adamson & Gast, 1997; Keller & Staudt, 2006; Birdi, 2003, 2017, 2020; Sing et al., 1985).

Type I gas adsorption isotherm: The most commonly applied adsorption model for gas reservoirs (for example: shale reservoirs) (Yu et al., 2014; Birdi, 2016, 2020) is the classic *Langmuir isoth*erm (Type I) (Adamson

**TYPE I
ISOTHERM**

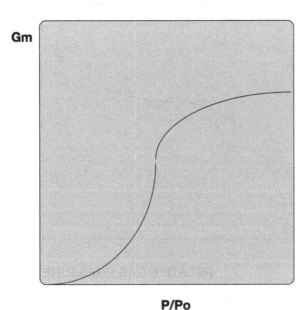

FIGURE 2.13   Gas – solid adsorption isotherms (amount of gas adsorbed versus gas pressure) (Type I).

and Gast, 1957; Birdi, 2020), which is based on the assumption that there is a dynamic equilibrium at constant temperature and pressure between adsorbed and non-adsorbed gas. Also, it is assumed that there is only a single layer of gas molecules adsorbed on the solid surface.

Further, gas – adsorption data showed that (Figure 2.1) the monolayer formation on the solid surface, corresponds with inflection in the isotherm in Figure 2.2a. The monolayer structure is related to the calculation based on the amount of gas adsorbed and the area/molecule data. In the adsorption of this type (generally called: The Langmuir isotherm) the data can be analyzed by the relation:

$$v(p) = (v_L p)/(p + p_L) \tag{2.21}$$

where $v(p)$ is the gas volume of adsorption at pressure p; v_L is the Langmuir volume.

The quantity v_L corresponds to the maximum gas volume of adsorption at infinite pressure; p_L is Langmuir pressure, which is the pressure corresponding to one-half Langmuir volume. It is assumed that there exists instantaneous equilibrium of the sorbing surface and the storage in the pore space is assumed to be established (Gao et al., 1994; Keller & Staudt, 2006).

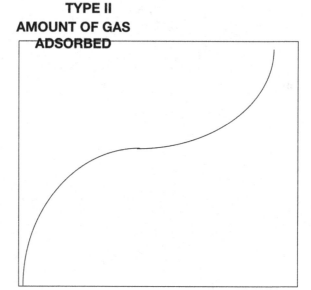

RELATIVE GAS PRESSURE

FIGURE 2.14 Gas – solid adsorption isotherms (amount of gas adsorbed versus gas pressure) (Type II).

It is also found that in some systems, at high gas pressures, the gas adsorbed on the solid surfaces forms **multi-molecular layers** (Figure 2.1). This is indicative of the effect of high gas pressure on the characteristics of adsorption. In other words, the Langmuir isotherm will not be an appropriate approximation of the amount of gas adsorbed (Figure 2.17).

Type II gas adsorption isotherm: This type of adsorption isotherm corresponds to multilayer sorption of gas, and the gas adsorption isotherm of Type II should be expected to be valid (Figure 2.7). After the monolayer gas is formed, as pressure increases, multi layer structures are formed (Figure 2.2). Experiments show that the Type II isotherm is often observed in a non-porous or a macroporous material (Freeman et al. theory in the Journal of the American Chemical Society; Sing et al., 1985).

The BET isotherm model is a generalization of the Langmuir model to multiple adsorbed layers (as shown in Figure 2.2). The BET model of adsorption assumes a homogeneous surface of the solid, and there exists no lateral interaction between the adsorbed molecules. The equation of BET also assumes that the uppermost layer is in equilibrium with the gas phase.

The data plot of $p_{gas}/v \, (p_o - p_{gas})$ versus p_{gas}/p_o (generally a straight line) which gives an estimate of the value of $1/v_m \, C_o$ at the intercept) and the slope is equal to the quantity:

$$(C_c - 1)/v_m \, C_o \tag{2.22}$$

**TYPE III
Gas Adsorption
On Solid**

Amount Adsorbed

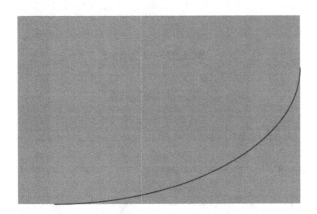

Relative Pressure

FIGURE 2.15   Gas – solid adsorption isotherms (amount of gas adsorbed versus gas pressure) (Type III).

Where C_c is a constant.

From the analyses of these data and from the magnitude of vm, the specific surface area can be estimated.

The standard BET isotherm theory assumes that the number of adsorption layers is infinite.

Type III gas adsorption isotherm: This type of adsorption isotherm is only observed in some rare adsorbent materials (Figure 2.18). One instance is the adsorption of nitrogen on ice (Adamson & Gast, 1997). There is some indication of multi-layer formation (Figure 2.18).

In this type of adsorption, the process of gas does not form a monolayer, but instead clusters are known to be formed.

Type IV gas adsorption isotherm: This type of isotherm is indicative of monolayer/bilayer/multi-layer structures (Figure 2.2). This type is also called capillary condensation type. This type is typical as found in porous solids (Figures 2.19–2.21).

Type I & II
Molecular states

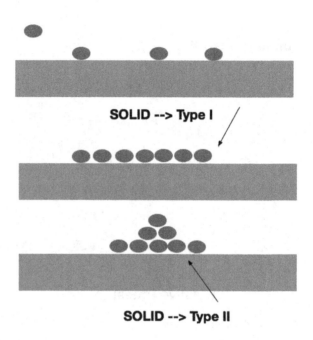

FIGURE 2.16   Gas structures corresponding to different isotherms (Types I and II).

Type V gas adsorption isotherm: These isotherms are generally observed with multi-layer adsorption systems.

Type VI adsorption isotherm: The isotherm (the stepped) of Type VI, which is relatively rare, is also reported in the literature. Type IV (Figure 2.2) and V isotherms typically exhibit a hysteresis loop, which is characteristic of porous systems, involving capillary condensation.

Obviously, it is found that the process of gas adsorption on a solid is described through isotherms, i.e., through the functions connecting the amount of adsorbate (i.e. gas atom/molecule) taken up by the adsorbent (solid) (or the change of any other physical parameter related to the adsorption of matter) with the adsorptive equilibrium pressure p, the temperature T, and all other parameters being constant. Below the critical temperature the pressure is appropriately normalized to the saturation vapor pressure p°, and the adsorbed amounts are so referred to the dimensionless relative pressure, p/po.

The quantity surface **fractional coverage** θ of the adsorbate, at a given equilibrium pressure p, is defined as the ratio of N_s surface sites occupied by the adsorbate

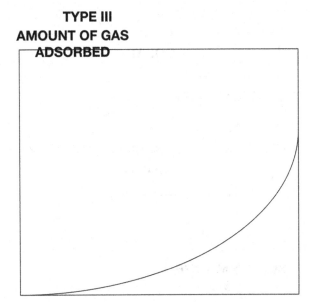

TYPE III
AMOUNT OF GAS
ADSORBED

RELATIVE GAS PRESSURE

FIGURE 2.17   Gas – solid adsorption isotherms (amount of gas adsorbed versus gas pressure (Type III).

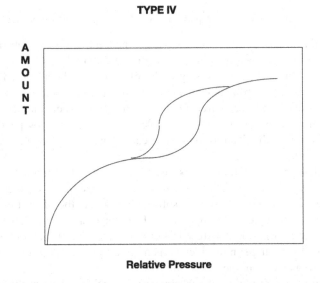

TYPE IV

A
M
O
U
N
T

Relative Pressure

FIGURE 2.18   Gas – solid adsorption isotherms (amount of gas adsorbed versus gas pressure) (Type IV).

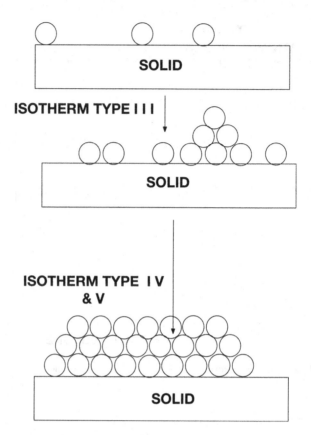

ISOTHERM TYPE I I I

ISOTHERM TYPE I V
& V

FIGURE 2.19 Gas structures corresponding to different isotherms (Types III, IV, and V).

over the total available adsorption sites N_T, i.e. the total number of substrate surface sites which are active towards the given adsorptive system.

In this case, the more specific step is the **first** layer of the adsorbed phase. The adsorption energy arises from either *chemisorption* or *physisorption*, or both, according to the nature of the forces governing the adsorbate/adsorbent interactions. On the other hand, the **second** layer of the gas molecules is originated by physical forces, similar to the forces that lead to the non- ideal behavior of gases and eventually to the condensation to the liquid.

The **multi-layer** adsorption of gas layers is expected to approach a *liquid-like* phase structure on the solid. This means that gas molecules are packed close and will thus expected to behave like a liquid or solid (solid-like). In the adsorption process, as the number of N_S occupied sites approaches the number of total available sites N, the adsorbate monolayer is complete ($\theta = 1$). For any further adsorption of gas, the latter will have to adsorb on gas molecules (which are adsorbed). The adsorption energy will thus show a change in the isotherm.

The (Figure 2.2) formation of subsequent layers of adsorbate at the surface of a solid sample is schematically illustrated. The process of gas adsorption is dependent

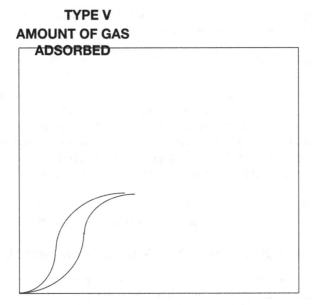

FIGURE 2.20   Gas – solid adsorption isotherms (amount of gas adsorbed versus gas pressure) (Type V).

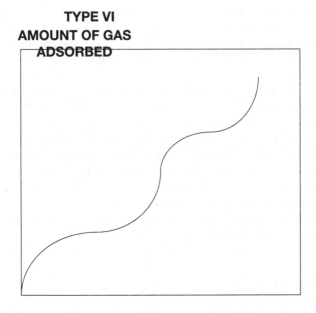

FIGURE 2.21   Gas – solid adsorption isotherms (amount of gas adsorbed versus gas pressure) (Type VI).

on various experimental parameters. The amount of gas taken up by a solid surface depends upon the solid and the gas nature:

- , the pressure *p* of the gas
- and the temperature T.
- The degree of adsorption being proportional to the mass m and the surface area A of the sample, the adsorbed amounts (often expressed as mass or volume STP of the gas) are properly normalized either to the unit mass or to the unit surface area. Here, in view of describing the process at molecular details, the adsorbed amounts n_{ads} are appropriately expressed as adsorbate moles (or molecules) per either unit mass or unit surface area of the adsorbent.

For example: The adsorption data for CO (at T = 303°K) on Na– and K–MFI zeolites have been investigated (both by volumetric and calorimetric methods) (Birdi, 2020).

2.10   SOLID SURFACE STRUCTURES AND GAS ADSORPTION

In any system, when a molecule (or an atom) from the gas-phase approaches a solid, it is more or less attracted by the atoms exposed at the solid surface, according to the nature of both the gas molecule and the solid material (Chapter 3).

The molecular arrangement of the solid phase has different characteristics based on the single crystal structure. For example, a crystalline solid is described through the periodic infinite repetition of an elemental pattern (unit cell). However, the symmetry of the periodic repetition of the unit cell terminates at the surface, thus creating asymmetrical force interaction. This molecular description has been studied in molecular detail by suitable atomic microscopes.

Its structure depends on the cleavage of the crystal, the chemical (either ionic or covalent) nature of the solid, and the origin of the surface (either chemical or mechanical).

The surface atoms arrangement depends on the plane preferentially exposed during the formation of the surface, according to the partition conditions of the real material (either in the single crystal form or as nano-sized powder). If no major reconstruction processes are required in order to minimize the surface atoms energy, and if no structural/compositional defects are present, an ideal perfect homogeneous surface is obtained which can be properly represented by cutting a slab of the solid structure. Such an ideal perfect homogeneous surface is very rarely encountered, unless especially prepared for surface science studies.

Real solid surfaces (mostly in the case of finely divided, nano-metric sized solids) are made up of a combination of:

flat regions (terraces): -_-__-_-__-_---__--
structural defects, such as:
  steps: ---__---__—
  kinks: - - - -
  corners: /\\\\\\\\\\\\\
  edges: ---|||||---
  point defects: --- --- ---

vacancies of ions/atoms in the solid: $+$+++ +++ --- ---.

These surface characteristics thus suggest that gas adsorption will also be dependent on these parameters (as also found from experiments). Compositional defects may contribute to the "imperfections" of the solid surface.

These include a variety of oxidation states of the atoms constituting the solid and/ or a variety of heteroatom present either as impurities, or especially introduced in order to modify the physico-chemical properties of the surface.

During the past few decades, with the advent of high resolution electron microscopes and other atomic microscopes it has become possible to image surfaces of solids with atomic details (at nano-meter scale) (Birdi, 2017, 2020).

For example, structural defects at the surface have been imaged at the molecular scale by high resolution transmission electron microscopy (HR-TEM), as found for monoclinic ZrO_2 nano-crystals, which terminate with structural defects as steps, kinks edges, and corners (Bolis et al., 1989). The surface defects are found to give rise to valence unsaturation thus creating a surface, which has highly reactive sites for gas adsorption.

Porous Solid Materials: The surface properties of a solid are dependent on various factors. The most important one arises from the size of solid particles. Finely divided solids possess not only a geometrical surface, as defined by the different planes exposed by the solid, but also an internal surface due to the primary particle aggregation. This leads to "pores of different sizes", according to both the nature of the solid and origin of the surface. Experiments show that these pores may be circular; square, or other shape. The porous solids (Figure 2.2):

/S/ /S/ /S/ /S/ /S/ /S/ /S

/S/ /S/ /S/ /S/ /S/ /S/ /S

/S/ /S/ /S/ /S/ /S/ /S/ /S

where pores are indicated as / /.

The gas (G) molecules will adsorb in the pores as indicated:

/S/G /S/G /S/G /S/G /S/G /S/G /S

/S/ G/S/G /S/G /S/G /S/G /S/G /S

/S/ G/S/G /S/G /S/G /S/G /S/G /S

It is thus obvious that much larger amounts of gas will be expected to adsorb in porous solids than non-porous materials. This process is also called absorption in solids. The size of pores is designated as the average value of the width, w (Gregg and Sing, 1982).

The width, w_p, gives either the diameter of a cylindrical pore or the distance between the sides of a slit-shaped pore. The smallest pores, with the range of width $w_p < 20$ Å (2 nm), are called *micropores*. The mesopores are in the range of a width 20 Å $\leq w \leq 500$ Å (2 and 50 nm). The largest pores, in the range of width $w_p > 500$ Å (50 nm), are called *macropores* (Birdi, 2017, 2020).

The shapes of pores will vary in geometric size and shape (e.g. circular, square, and triangular). The capillary forces (Appendix A) in these pores will thus depend both on the diameter and on the shape. In general, most solid adsorbents exhibit (for example: like charcoal and silico-aluminates) irregular pores with widely variable diameters in a normal shape. Another important criterion for porous substances is as regards the connectivity of individual pores. The degree of connectivity is sometimes observed in the gas adsorption isotherms. Some other adsorbents, conversely, such as zeolites and clay minerals) are entirely micro- or meso-porous, respectively. In other words, the porosity in these materials is found not to be related to the primary particle aggregation but it is an intrinsic structural property of the solid material (Rabo, 1981; Birdi, 2016, 2020).

2.11 THERMODYNAMICS (BASICS) OF GAS ADSORPTION ON SOLID

In the system: gas phase – solid phase, the gas molecules are different as regards the kinetic movement. From the equation relating free energy (ΔG) and entropy (ΔS) of any system:

$$\Delta G = \Delta H - T \, \Delta S \tag{2.23}$$

The basic description as regards the thermodynamics of gas adsorption on solid surfaces has been reported in the literature (Chattoraj & Birdi, 1984; Adamson & Gast, 1997; Keller & Staudt, 2006; Birdi, 2020).

One finds that the quantity enthalpy (ΔH_{ads}) of adsorption can be measured by using a suitable calorimeter. The kinetic theory of gases shows that molecules are moving in space. Gas molecules will thus be expected to lose most of the kinetic movements after adsorption on a solid. As mentioned earlier, the gas molecules are thus at a lower entropy after adsorption.

The adsorption of a gas at a solid surface is found to be an exothermic process (enthalpy of adsorption). This is expected by the thermodynamic condition for a spontaneous process:

$$\Delta G_{ads} = \Delta H_{ads} - T \, \Delta S_{ads} < 0 \tag{2.24}$$

In fact, adsorption being necessarily accompanied by a **decrease** in entropy ($\Delta S_{ads} < 0$) in that the degrees of freedom of the molecules in the adsorbed state are lower than in the gaseous state.

Experiments show that the value of ΔH_{ads}, (the enthalpy change of gas adsorption) is expected to be negative (Birdi, 2020).

2.12 DETERMINATION OF HEAT (ENTHALPY) OF GAS ADSORPTION ON SOLIDS (FROM INDIRECT NON-CALORIMETRIC METHODS)

The quantity **enthalpy** of any system can be measured by various methods. The direct method is where a suitable calorimeter is used (Appendix B). Many commercial calorimeters (micro-calorimeters) are available with varying sensitivity characteristics. The sensitivity of the calorimeter is selected with reference to the system to be investigated (i.e. the magnitude of the change in temperatures of adsorption process (Appendix B).

The magnitude of the quantity enthalpy of any system can also be estimated from the change in free energy with temperature. The quantity enthalpy of any system can also be obtained by other procedures than the direct calorimetric, i.e. from the change of free energy with temperature. This equation is called the Clausius – Clapeyron equation (Keller & Staudt, 2006; Adamson & Gast, 1997; Birdi, 2020):

$$h_{ads} = R \, T^2 \left(d \ln pg \, / \, d \, T \right) N_s,\qquad (2.25)$$

$$= \Delta H_{ads}\qquad (2.26)$$

p_{g1} and p_{g2} values at T_1 and T_2, for a given constant coverage θ:

$$\ln\left(p_{g1}/p_{g2}\right) = q_{s1} \, / \, R\left((1/T_2) - (1/T_1)\right)\qquad (2.27)$$

As an example: the data of N_2 adsorption on H–ZSM5 were investigated (in the 104–183 K temperature range). The adsorption process was studied by using the change upon adsorption of the absorbance intensity of the νOH stretching band at $3{,}616\,cm^{-1}$, as related to the Bronsted acidic site $Si(OH)^+Al^-$. From these data, the adsorption enthalpy was estimated to be (Birdi, 2020):

$$\Delta H^\circ_{ads} = -19.7 \pm 0.5 \text{ kJ mol}^{-1} \text{(calorimetric heats (}h_{ads}\qquad (2.28)$$

$$= 19 \text{ kJmol}^{-1}\text{) measured at } T = 195 \text{ K for } N_2 \text{ on } H-ZSM5\qquad (2.29)$$

For example: Measurements of adsorption of gas NH3 on a solid H–ZSM5 zeolite were reported (Armandi et al., 2010):

$$\Delta H^\circ_{ads} = -128 \pm 5 \text{ kJ mol}^{-1},\qquad (2.30)$$

(calorimetric heat of adsorption ($q_o \approx 120\,kJ \text{ mol}^{-1}$).

These data thus show good agreements between the two methods for determining the enthalpy of gas adsorption on different solids.

2.13   ENTROPY OF GAS – SOLID ADSORPTION (S_{ADS})

The gas molecules occupy almost 1,000 times more volume than the same molecules in a liquid or solid phase. This means that the molecules in the gas phase have higher entropy than when these are in the adsorbed state (on a solid surface). It is thus important to consider the entropy, S_{ads}, of gas - solid adsorption processes. The entropy change ΔS_{ads} related to the gas adsorption in an ideal system case was estimated, by using the statistical mechanics (rotor/harmonic oscillator) formula (Quarrie, 1986).

As a typical example: The argon (Ar) gas atom, which adsorbs on a solid surface, is known to loose entropy from the gas phase: the translation entropy, S_{tr}, of the solid, which is fixed in the space, is taken as zero, whereas the free Ar atoms, before interacting with the solid surface, possess a translational entropy Str, which amounts to 150 and 170 J mol^{-1} K^{-1} at T = 100° and 298° K, respectively, at p_{Ar} = 100 Torr.

It has been suggested that the adsorbed gas atom/molecule (GGG) will be expected to possess lesser degree of translational free energy. After adsorption, it will exhibit some vibrational energy.

GAS PHASE......G (KINETIC MOVEMENT IN THREE DIMENSIONS

ADSORBED STATE.....G (VIBRATIONAL MOVEMENT IN TWO DIMENSIONS)

Assuming that the vibrational frequency ≈ 100 cm^{-1}, the adsorbed atoms, then one finds that the magnitude of the vibrational entropy is S_v = 18 and 43 J mol^{-1} K^{-1} at p_{Ar} = 100 Torr and at T = 100° and 298° K, respectively (Figure 2.22).

As expected, the adsorbed atom entropy is much lower than the free gas atom entropy, and the entropy change $\Delta S_a = (S_v - S_{tr})$ is, in all cases, negative: $\Delta S_{ads} = -132$ J mol^{-1} K^{-1} at T = 100° K and $\Delta S_a^\circ = -127$ J mol^{-1} K^{-1} at T = 298° K.

This means that in a spontaneous process, which requires a *negative* free energy change ($-\Delta G$), the enthalpy of adsorption must be negative in order to compensate the loss of entropy. In other words, the process must be exothermic of an amount of heat evolved at least as high as the decrease of the T $\Delta_{ads}S$ quantity.

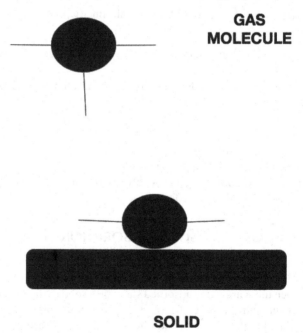

GAS MOLECULE

SOLID

FIGURE 2.22 The different kinetic movements of gas molecule in: gas phase; adsorbed on a solid (each unit -of degree of gas molecule movement is depicted by an arrow).

The integral molar entropy of adsorption is the difference between the entropy of an adsorbed molecule and the entropy of the adsorptive in the ideal gas state, at given p and T. It is a mean integral quantity taken over the whole amount adsorbed and it is characteristic of a given state of equilibrium. This is distinguished by the standard integral molar entropy of adsorption, which is the entropy of one adsorbed mole with respect to the entropy of the adsorptive in the ideal gas state at the same T, but under standard pressure (Garrone et al., 1999; Savitz et al., 1999; Birdi, 2020).

2.14 ADVANCED EXPERIMENTAL PROCEDURES OF GAS ADSORPTION ON SOLIDS: STM (SCANNING TUNNELING MICROSCOPE)

The mechanism of adsorption of gas on solid surfaces requires the information about the molecular packing of the adsorbate. This information, at the molecular scale, is useful in the characterization of the adsorbed gas layer (layers) in many ways. One needs to know the number of sites/unit surface area of a solid. This information thus provides the data about the packing geometry of the gas on the solid.

The gas adsorption studies on the adsorption of CO_{2gas} on solid surfaces were reported.

It was found that that the 1′t ME peak in TPD corresponds to four CO_2 molecules per unit cell.

As expected, the C1-s peak area from CO_2 is twice that of the formate species, which has a saturation coverage of two molecules per ° unit cell due to its bidentate binding configuration.

In another study (Birdi, 2020), the molecular details of molecules on surfaces can be investigated by using scanning probe microscopes (SPM). The investigation of atomic scale and microscopic scale structure of surfaces of solids is important.

The arrangement of the CO_2 molecules as adsorbed on the $Fe_3O_4(001)$ solid surface was performed by STM experiments. Empty state images of the as-prepared surface exhibit the characteristic undulating rows of protrusions related to fivefold-coordinated Fe^{3+} cations within a distorted surface layer.

These STM image analyses have shown much molecular detail, as regards the adsorption process.

The STM image of the unit cell showed different degrees of greyness depending on the degree of adsorption (Birdi, 2020). Data were obtained from the STM image acquired following the saturation exposure of CO_2 (at a sample temperature of 82 K).

Under the experimental conditions chosen, i.e. temperature is within the 2~d ML resorption peak, so only the first ML molecules should be present on the surface when the STM experiment is conducted.

The scanning microscopic data showed that the position of the bright and dark pairs relative to the underlying substrate in To and 7d was determined by watching the formation of the over-layer by dosing CO_2 directly into STM whilst scanning).

The location of CO_2 sites was found to be related to the degree of bright and dark areas in the scans.

By aligning the **before** and after **images** to surface defects it was found to assign the location of CO_2-molecules (as protrusions) to the surface. Furthermore, it was found that the images of the surface with a sub-monolayer CO_2 coverage showed *islands*.

STM data were also obtained where adsorption was in the initial stage (i.e. low coverage).

The STM analyses of a clean surface showed several defects, which were identified as surface hydro-surface groups (Birdi, 2020). The lattice defect is related to the presence of an additional Fe atom in the surface.

The incorporated Fe and APDB defects were recently shown to contain Fe^{2+} cations, and to be active sites for methanol adsorption.

Here we observe a preferential adsorption of CO_{2gas}, with bright protrusions appearing at the position of the defects while with the STM, but no similar events on the defect-free surface in-between. This shows that CO_2, a Lewis acid-like-methanol, interacts more strongly with the Fe^{2+} sites associated with these defects than with the regular type.

From the scanning tunneling microscope (STM), the STM images, quantitative TPD measurements, and spectroscopic data were presented here. From this, it was clear that the CO_2 unit cell was present, and each molecule is associated with one surface Fe cation. The clear separation of desorption from F and Fe^3 -related sites means that CO_2 can be a useful probe of the relative density of such sites on magnetite surfaces.

2.15   GAS – LIQUID (SOLUTION) ABSORPTION ESSENTIALS

In some instances, the gas-absorption in fluids is determined to be feasible.

In an interface between gas – liquid, there exists equilibrium between the two phases (CEE).

Any gas – liquid (or fluid) in reaction is dependent on the physical characteristics of the system.
The solubility of different gases in liquids is known to vary. Some gas molecules dissolve sparingly in liquid, while other gas molecules may dissolve/interact with high solubility.

In the present case: atmosphere and water in oceans/lakes/rivers interact at the atmosphere-olr interfaces. Since more than 70% of surface of earth is oceans/lakes/rivers (olr), this gas-liquid interface is known to be significantly large. Furthermore,

---

EXAMPLE

>ATMOSPHERE- OCEANS/LAKES/RIVERS (OLR)
>(SINCE CO_{2gas} IS THE ONLY COMPONENT WHICH CAN DISSOLVE IN OLR>
>$CO_{2,air}$: $CO_{2,aq}$

---

the depth of big oceans is known to reach over 5 km. This means that most of the earth climate characteristics depend largely on the latter.

In fact, the atmosphere – ocean interface is known to be the largest chemical zone. The latter is also accepted to play the major role in any earth climate phenomena (Kemp et al., 2022; Birdi, 2020).
This gives rise to many important consequences in the consideration of mass balance of carbon dioxide. The atmosphere – oceans interface is thus a natural carbon sink. In fact, it is recognized that due to its large area and volume, the interface constitutes as the largest interfacial phenomena (Kemp et al., 2022).

In the case of carbon *recycling* (and CCRS) of carbon dioxide, CO_2, various fluids have been used. It is known that if a gas is bubbled into a suitable fluid, the latter will dissolve/react to varying degrees. Gas absorption in fluids has been studied extensively, since it is an important phenomena in everyday life (technical processes and biological phenomena). Some of the most significant examples are: the treatment of pollutants removal from flue gases. Carbon dioxide (CO_2) is known to be soluble in water (Chapter 2; Appendix A). Hence, if CO_{2gas} is bubbled in water it will absorb, and is known to form carbonic acid, H_2CO_3. This is the primary interaction between the two molecules, e.g. water and carbon dioxide.

<Solubility of CO_{2gas} in Water (aq) [CO_{2aq}]:

$$\mu_{CO2gas} = \mu_{CO2aq} \qquad (2.31)$$

The equilibrium constant, $K_{CO2gas/aq}$, is:

$$K_{CO2gas/aq} = \left[CO_{2aq} \right] / \left[CO_{2gas} \right] \qquad (2.32)$$

where
μ_{CO2gas} and μ_{CO2aq} are chemical potentials of CO_2 in air and water, respectively; $K_{CO2gas/aq}$ is the equilibrium constant; (CO_{2aq}) and (CO_{2gas}) are concentrations of CO_2 in water and air, respectively (ideal conditions). In all equilibrium states, one must use activities (Chattoraj & Birdi, 1984).

However, if CO_{2gas} is bubbled in an aqueous solution such as NaOH (sodium hydroxide), then it will react to form Na_2CO_3. This will thus give rise to enhanced absorption. The process of gas capture by absorption in general is described as (Figure 2.23):

GAS BUBBLES – FLUID (SOLVENT) – GAS IN SOLUTION

In most systems the gas interacts with some components in the fluid (solution) (Yu et al., 2012; Hinkov et al. 2016; Birdi, 2020) (Figure 2.23).

This reaction gives rise to enhanced absorption of the gas.

For example: The latter process is based on the absorption (selectively) of CO_{2gas} in a solvent (solution). CO_{2gas} is extracted from the second stage of scrubber (Figure 2.23),

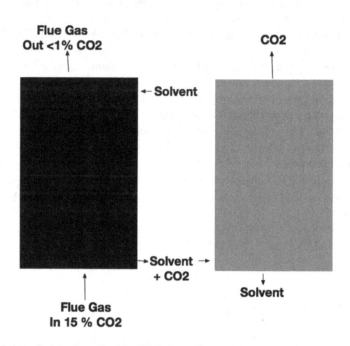

FIGURE 2.23   Gas (carbon dioxide: CO_2) absorption process.

where CO_{2gas} (ca. 99%) is recovered. This gives rise to a specific absorption. For example: CO_{2aq} in a solution of NaOH will form sodium carbonate salt (Na_2CO_3). This would thus enhance the absorption of CO_{2gas} in solution.

The concentration of CO_{2aq} will be almost the same as the concentration of NaOH. This arises from the fact that equivalent amounts of species interact as 1:1.

In the literature one finds many systems where aqueous solutions of amines have been used to capture CO_{2gas} (Chapter 4).

In another case: Gas is injected (a suitable scrubber) into a mixture of water + organic substance (non-miscible) (alcohol, alkane, etc.). The water/organic liquid system ($liquid_1/liquid_2$) creates a two-phase system. These gas - $liquid_{water}$ - $liquid_{immiscible}$ systems have gained interest in the past decade due to the introduction of homogeneous biphasic catalysis in various reaction systems, e.g. hydro- formylation, carbonylation, hydrogenation, and oligomerization (Birdi, 2020). The main advantage of these systems over catalysis in one phase is the easy separation of the catalyst and the reactants or products.

Gas-liquid$_1$-liquid$_2$ (where: liquid$_1$ is water (solvent) and liquid$_2$ is a solute) systems are further encountered in reaction systems which inherently consist of three phases due to two (or more) immiscible reactants, reaction products, or catalysts (Birdi, 2020).

Gas-liquid$_1$-liquid$_2$ systems are developed to add increased absorption characteristics and increased specificity. The approach in this kind of process is to add an additional (inert) liquid phase, in order to increase the mass transfer rate. This latter approach is also been applied to some biochemical applications (Birdi, 2020). However, the addition of a second liquid phase can also retard the gas - liquid mass transfer.

Some specific CO_{2gas} absorption studies are given in Chapter 4.

Gas absorption in fluids has been investigated for many decades (Birdi, 2020). The absorption of oxygen was reported in dispersions of water organic fluids (such as kerosene and toluene). There are also reported absorption studies on the effect of a second immiscible liquid on the gas - liquid interfacial area (Birdi, 2020) which used a fast reaction, CO_2 - NaOH system, to study the influence of 2-ethyl-hexanol on the specific gas - liquid interfacial area. It was found that the interfacial area increased due to a decreased degree of bubble coalescence.

2.16   ABSORPTION OF CO$_2$ IN AQUEOUS SOLUTIONS

Air consists (current equilibrium composition) of 80% oxygen + 18% nitrogen + 420 ppm (0.042%) CO_2 + minor amounts of other gases. Nitrogen/oxygen are almost insoluble in water. On the other hand, carbon dioxide reacts with water, H_2O, and forms carbonic acid (weak acid): H_2CO_3. It is known that besides low concentration of carbon dioxide in air, it can absorb into aqueous media (such as: oceans/lakes/rivers/rain drops; lung function). This chemical property of CO_{2gas} has been the most significant characteristic.

As an absorption/reaction system, carbon dioxide (gas) absorption in a 0.5 M potassium carbonate/0.5 M potassium bicarbonate buffer solution was used. The carbonate/bicarbonate buffer solution is the continuous liquid in these studies. As dispersed liquid phase several different organic liquids were added (such as: toluene, n-dodecane, n-heptane, and 1-octanol).

The absorption of CO_2 (gas) in a aqueous solution has been analyzed (Cents et al., 2001). In the absorption in the bulk liquid phase, the following reactions are taking place:

System:::CO_2 + POTASSIUM CARBONATE/BICARBONATE BUFFER (0.5 M)

$$CO_2 + H_2O = H_2CO_3 = HCO_3^- + H^+ \qquad (2.33)$$

The equilibrium constant for the carbonate reaction, K_{car}:

$$K_{car} = \left(k_f^{-1} \right) / \left(k_r^{+1} \right) \qquad (2.34)$$

With forward/reverse rate ratio of k_f^{+1}/k_r^{-1}).

There is another reaction which follows:

$$CO_3^{-2} + H^+ = HCO_3^+ \qquad (2.35)$$

It is known that this reaction can be catalyzed by different additives (such as: hypochlorite; arsenide). At high pH =10, as present in this buffer, there are also hydroxyl ions (OH⁻). CO_2 reacts with OH⁻:

$$CO_2 + OH^- = HCO_3^- \qquad (2.36)$$

In this relation, the equilibrium concentration of carbon dioxide is dependent on the **carbonate** and **bicarbonate** concentrations (Birdi, 2020).

3 Essential Surface Chemistry Aspects for Climate Phenomena

Surface Chemistry (Liquids and Solids) (Climate Change Aspects)

3.1 INTRODUCTION

It is important to add useful descriptions necessary for analyzing the essential surface chemistry principles for climate carbon recycling phenomena on earth.

It has been recognized (Birdi, 2020) that as regards the control/recycling/mitigation of one of the major GHGs, i.e., carbon dioxide ($CO_{2,gas}$), surface chemistry principles are of one of the major property.

The GHG effect of carbon dioxide (CO_{2gas}) is only present from the gaseous state. Furthermore, the GHG effect in the environment has existed since photosynthesis was observed on earth, billions of years ago. In other words, both GHG and photosynthesis effects have varied over these geological times. The complexity of the latter phenomena is also obvious, and recognized in climate model analysis.

At the interface of atmosphere-oceans: It is known that carbon dioxide (CO_{2gas}) dissolves in water (as: CO_{2aq}) (e.g. oceans/lakes/rivers rain drops). The GHG effect is absent in the case of CO_{2aq} dissolved in water (e.g. oceans/lakes/rivers). In addition, it is found that the mass balance between these phases is at pseudo equilibrium (Appendix A). This arises mainly from the fact that gas – liquid mixing is inadequate in oceans.

Furthermore, the purpose of this chapter is to provide some basic and essential information about different terms related to surface chemistry as used throughout this book. It is useful to mention that the current literature has indicated the important role of surface chemistry in the investigation of climate aspects (such as: any man-made technological advancements (increasing usage of fossil fuels)). It is also essential to mention that surface chemistry principles are important factors in the consideration of the climate and environmental conditions surrounding the earth.

It is important to describe the different states (phases) of matter as found in the universe (Chapter 3). All the matter that exists around the universe consists of different distinct phases:

DOI: 10.1201/9781003300250-3

- gas,
- liquid, and
- solid phases.

All these phases exhibit specific characteristics.

The gas phase: its molecules fill a container regardless of the shape of the container (Figure 3.1). This is because the gas molecules are free to move (without breaking any bonds) and thus fill the whole volume of any container.

The liquid phase also fills a container corresponding to its volume (contrary to the gas) regardless of shape of the container (Figure 3.1).

A *solid* phase, however, keeps its shape, no matter what container it is placed in (Figure 3.1). The molecules in a solid are strongly bonded to each other and cannot move unhindered. The distance between molecules in the solid phase is generally 10% shorter than that in the liquid phase. In liquids, the distance is approximately ten times shorter than that in the gas phase (Appendix A). However, the ice phase is observed to be an exception. The ice-water system shows anomalous behavior.

In the *ice* phase the density is about 10% *lower* than that in water at 0°C. This being the reason that iceberg floats on the surface of oceans. This anomalous behavior

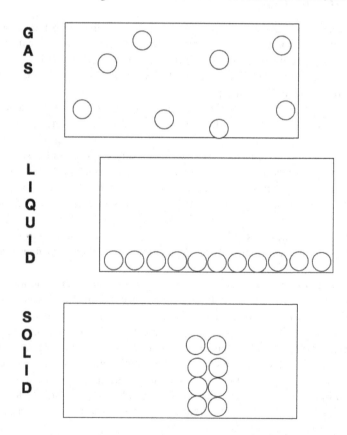

FIGURE 3.1 Gas-Liquid-Solid states of matter. (See text).

of the water – ice system has also implications on climate change and oceans. In other words, when ice-berg melts, there is no change in volume of the surroundings (because water is heavier than ice-berg). Furthermore, the open structure of ice is known to capture other suitable molecules. This ice-molecule complex is called clathrate (Chapter 4).

These considerations are based on macroscopic dimensions. The state of molecules is rather different, as follows. These are called **bulk phases**. In the bulk phase, each atom/molecule is surrounded by *symmetrical* neighboring atoms/molecules: (dark OO)

OOOOOOOOOOOOO
OOOOOOOOOOOOO
OOOOOOOOOOOOO

The molecule (**O**) is surrounded by the same kinds of neighbors. Thus it interacts with equivalent kinds of forces (symmetrical) and is stable (Figure 3.2).

The schematic view of intramolecular ratio of near neighborhoods is about 2:4. This ratio is supported by other estimates (see Chapter 3). The state of forces acting on a molecule inside a bulk phase and surface is found to be asymmetrical. The surface

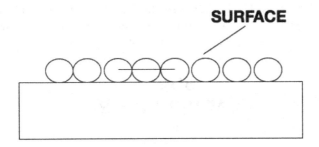

SURFACE

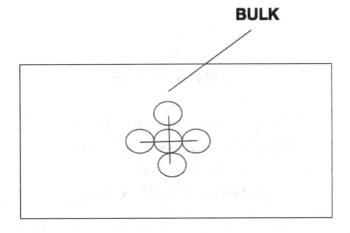

BULK

FIGURE 3.2   Molecular forces: (a) at surface; (b) in bulk phase. There are 2:4 neighbors.

chemistry is based on these asymmetrical forces arising from different sources (such as: van der Waals; electrical; hydrophilic-hydrophobic) (Birdi, 1999, 2020).

3.2 SURFACE TENSION OF LIQUIDS

The surface tension of a liquid can be measured directly. The accuracy of this has been found to very high (Birdi, 1982, 1989, 1999, 1993, 2003, 2009, 2916, 2020). The surface tension of solid surfaces, cannot be measured by any direct means. The latter has been estimated by indirect procedures.

3.2.1 INTRODUCTION

In this chapter the surface of the liquid phase is described, as related to the climate phenomenon (Birdi, 2020). Many natural phenomena are related to the characteristics of liquids (for example, water (rain, lakes, rivers, and oceans). The surface of oceans is found to play an extensive role in many important natural phenomena. The surface area of the oceans/lakes/rivers is much larger than the land area on earth (Figure 3.3).

Since the physical properties of terrestrial land/oceans are very different (e.g. density; heat reflection; heat absorption; other interactions; kinetics variables), there will be significant differences. The liquid surface forces are found to be very significant in many of these phenomena. It is important to mention that about 70% of the surface of Earth is covered by water (e.g., oceans, lakes, rivers). About 30% of the surface is

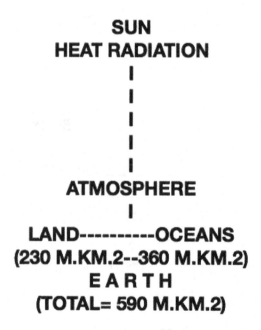

FIGURE 3.3 Atmosphere – water (oceans/lakes (rivers) surface (interface) interaction. Earth surface: land = 260 million km²; oceans = 560 million km².

covered by land (plains; deserts; Antarctic; forests). Furthermore, the great oceans can be over 5,000 m deep. This means that the volumes of oceans can be very significant. In fact, the atmosphere-ocean interface is the largest surface chemical phenomenon in nature. It is therefore recognized that the latter interface is very significant in the analysis of climate phenomena. Furthermore, carbon dioxide, CO_{2gas}, interacts with water in oceans to form hydrogen carbonic acid, H_2CO_3. The latter reaction gives rise to a variety of carbonated in oceans, which leads to the formation of shells and fisheries (Birdi, 2020). In addition, fishery (as food) is a major carbon dioxide (CO_{2gas}/CO_{2aq}) cycle.

Similarly, the importance of rivers and rain drops on various natural phenomena is very obvious. The effect of rain *drops* is also made up of dynamic interactions. It is therefore important to give a detailed introduction to the physico-chemical principles of the surface tension of liquids. The most fundamental characteristic of liquid surfaces is that they tend to contract to the smallest surface area in order to achieve the lowest free energy. Whereas gases have no definite shape or volume, completely filling a vessel of any size containing them, liquids have no definite shape but do have a definite volume, which means that a portion of the liquid takes up the shape of that part of a vessel containing it and occupies a definite volume, the free surface being plane except for capillary effects where it is in contact with the vessel. However, if the size of containers is very small, such as in porous materials, then the surface force (i.e., surface tension of liquids) becomes the dominant parameter. This is observed when one notices rain drops and soap films, in addition to many other systems. The cohesion forces present in liquids and solids and the condensation of vapors to liquid state indicate the presence of much larger intermolecular forces than gravitational forces. Furthermore, the dynamics of molecules at interfaces are important in a variety of areas, such as biochemistry, electrochemistry, and chromatography.

The degree of *sharpness* of a gas – liquid surface (or: liquid₁ – liquid₂ interface) has been the subject of much research in the literature. There is strong evidence that the change in density from liquid to vapor (by a factor of ca. 1,000) is exceedingly abrupt, that is, in terms of molecular dimensions. The surface of a liquid has been analyzed by light reflectance investigations, as described by Fresnel's law. Various investigators indeed found that the surface transition involves just one layer of molecules. In other words, when one mentions surfaces and investigations related to this part of a system, one actually mentions just a molecular layer. However, there exists one system which clearly shows the *one molecule thick* layer of surface as being the interface of a liquid: this is the monolayer studies of lipids spread on water (Birdi, 1999, 2020).

The surface thermodynamics of these monolayers is based on the uni-molecular layer at the interface, which thus confirms the thickness of the *surface*. The molecules of a liquid in the bulk phase are in a state of constant unordered motion like those of a gas, but they collide with one another much more frequently owing to the greater number of them in a given volume (Figure 3.4):

- **GAS PHASE** molecules in gas
- (*INTERMEDIATE PHASE*)
- **LIQUID SURFACE** surface molecules
- **BULK LIQUID PHASE** molecules inside liquid

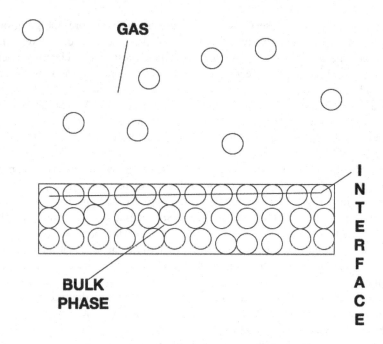

FIGURE 3.4   Liquid gas-phase/interface/liquid surface-phase/liquid bulk-phase

It is known that a gas molecule occupies (approximately) 1,000 times more volume than a molecule in the liquid (or solid) phase (Chapter 2). Further, the *intermediate* phase is only present between the gas phase and the liquid phase. Although one does not often think about how any interface behaves at equilibrium, the liquid surface demands special comment. The surface of a liquid is under constant agitation; there are a few things in nature presenting an appearance of more complete repose than a liquid surface at rest. However, according to the kinetic theory, the molecules are subject to much agitation. This is apparent if one considers the number of molecules that must evaporate each second from the surface in order to maintain the vapor pressure. At equilibrium the number of liquid molecules that evaporate into the gas phase is equal to the number of gas molecules that condense at the liquid surface (which will take place in the intermediate phase). The number of molecules hitting the liquid surface is considered to condense irreversibly. From the kinetic theory of gases, this quantity can be estimated as follows:

$$\text{mass} / \text{cm}^2 / \text{second} = \rho_G \left(k_B \, T \, / \, 2\pi \, m_m \right)^{0.5}$$
$$= 0.0583 \, p_{vap} \left(M_w T \right)$$

(3.1)

where k_B is the Boltzmann constant ($1.3805 \, 10^{-16}$ erg/deg), m_m is the mass of molecule, ρ_G is the density of the gas, and M_w is the molecular weight.

For example, in the case of water (at 20°C)

the vapor pressure of this liquid (water) is 17.5 mm,
which gives 0.25 g/sec/cm^2 (from equation 3.1).
This corresponds to 9×10^{21} molecules of water per second.

From consideration of the size of each water molecule one finds that there are ca. 10^{15} molecules, so that it can be concluded that the average life of each molecule in the surface is only about one eight-millionth of a second (1/8 10^{-6} sec). This must be compounded with the movement of the bulk water molecules toward the surface region. It thus becomes evident that there is an extremely violent agitation in the liquid surface. In fact, this turbulence may be considered analogous to the movement of the molecules in the gas phase. One observes this vividly in a cognac glass. The ethanol (in cognac) molecules evaporate and condense on the walls of the container (the boiling point of ethanol is lower than that of water).

In the case of interface between two immiscible liquids, due to the presence of interfacial tension, the interface tends to contract. The magnitude of interfacial tension is always lower than the surface tension of the liquid with the higher tension. The liquid – liquid interface has been investigated by specular reflection of X-rays to gain structural information at molecular (Angstrom ($Å = 10^{-8}$ cm $= 0.10$ nm) resolution (Adamson & Gast, 1997; Birdi, 2002, 2015, 2020).

The phenomena of capillary forces at curved surfaces: The term **capillarity** (a Latin word capillus: a hair) describes the rise of liquids in fine glass tubes. This observation added the study of curved surfaces of liquids.

The rise of fluids in a narrow capillary was related to the difference in pressure across the interface and the surface tension of the fluid (Figure 3.5):

$$\Delta P = 2 \yen \text{ (curvature)} = 2 \yen \text{€}(1/\text{radius of the curvature}) \tag{3.2}$$

This means that when a glass (or any other material) tube of a hair-fine diameter is dipped in water, the liquid meniscus will rise to the very same height. A fluid will *rise* in the capillary if it wets the surface, while it will decrease in height if it non-wets (like Hg in a glass capillary) (Adamson & Gast, 1997; Birdi, 2015, 2020).

The magnitude of rise of a liquid is rather large, that is, 3 cm if the bore is of 1 mm for water. This equation also explains what happens when liquid drops are formed at a faucet. Thus any curved liquid surface (Figure 3.6) exhibits capillary force. Although it may not be obvious here, but the capillary force can be very dominating in different processes (for example, the properties of a sponge or oil/gas recovery from a reservoir).

In Figure 3.3 it is found that the rise of liquid takes place due to ΔP only, since the liquid meniscus is curved. The curvature induces ΔP across the interface and liquid rises, corresponding to the magnitude of ΔP. At an equilibrium, the mass of fluid in the capillary is equal to ΔP.

The capillary force is a result of *curved* liquid surface. The fluid rise in the capillary is to balance by the rise/decrease of fluid height (Figure 3.7). In porous solid materials (such as sponges, soil, and gas/oil/shale reservoir), this capillary force thus becomes the most significant driving force (for example, oil/gas reservoirs, ground

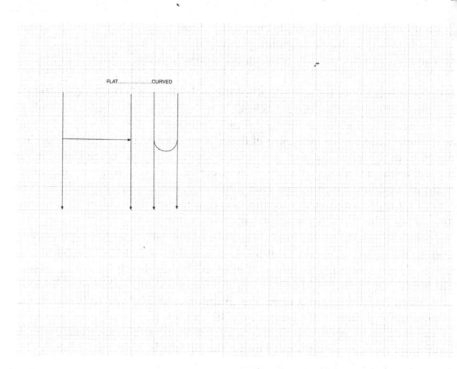

FIGURE 3.5 Liquid flat or curved surfaces: flat surface (DP = 0); curved surface (DP = /0).

water seepage, sponge, fabrics, etc.). The gas adsorption in porous solids exhibits capillary condensation phenomena (Adamson & Gast, 1997; Birdi, 1997, 2002).

The magnitude of capillary rise is *higher* in the smaller tubing than in the larger, since the magnitude of ΔP is higher in the former. This explains why it is difficult to recover oil from some reservoirs (such as shale reservoirs, which have very small pores). The same is found in the case of two bubbles or drops (Figure 3.8), where the smaller bubble or drop (due to lager ΔP) will coalescent with the larger bubble or drop.

The capillary phenomenon thus means that it will be expected to play an important role in all kinds of systems where liquid (with curved surface) is in contact with materials with pores or holes. In such systems the capillary forces will determine the characteristics of liquid – solid systems. Some of the most important ones are thus:

- all kinds of fluid flow inside solid matrices (ground water and oil recovery)
- fluid flow inside capillary (oil recovery, ground water flow, and blood flow)

It was recognized at a very early stage that only the forces from the molecules in the surface layer act on the capillary rise. An impressive example from everyday life is the flow of blood in all living species, which is dependent on capillary forces. The oil recovery technology in reservoirs (shale gas/oil) is similarly dependent on the capillary phenomena. The capillary forces become very dominating in latter systems. The most significant example is the shale gas oil reservoirs.

(A) FLAT SURFACE
(B) CURVED SURFACE

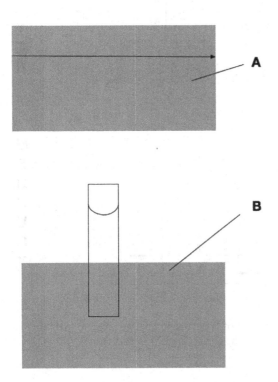

FIGURE 3.6 A flat surface ($\Delta P = 0$) and a curved liquid surface (ΔP = capillary pressure).

Furthermore, virtually all elements and chemical compounds are known to exist in three phases:

1. solid,
2. liquid,
3. and vapor phase.

A transition from one phase to another phase is accompanied by a change in temperature, pressure, density, or volume. This observation thus also suggests that because of the term ΔP, the chemical potential (with *curved* surfaces) will be different in systems with *flat* surfaces.

A typical molecular explanation can be useful to consider in regard to surface molecules. Molecules are small objects that behave as if of definite size and shape in all states of matter (e.g., gas [G], liquid [L], and solid [S]). The volume occupied by a molecule in the gas phase is generally 1000 times larger than the volume occupied by a molecule in the liquid phase, as follows:

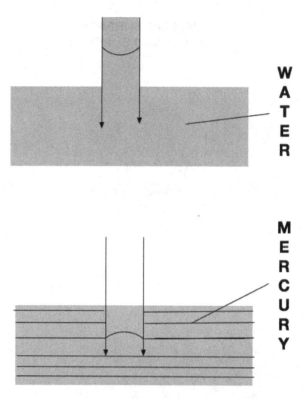

FIGURE 3.7 Capillary rise of liquid due to Laplace pressure (ΔP). (In the case of mercury (Hg), there is a fall because the contact angle is greater than 90°).

As shown above, the volume of one mole of a substance C, for example, water in the gas phase (at standard temperature and pressure), V_G (ca. 24,000 cc/mol), is some 1,000 times its volume in the liquid phase, V_L (molar volume of water = ca. 18 cc/mol). The distance between molecules, D, will be proportional to $V^{1/3}$ such that the distance in the gas phase, D_G, will be approximately 10 (= $1000^{1/3}$) times larger than that in the liquid phase, D_L.

The finite compressibility and the relatively high density, which characterize liquids in general, point to the existence of repulsive and attractive intermolecular forces. The same forces that are known to be present in the gaseous form of a substance may be imagined to play a role also in the liquid form. The mean speed of the molecules in the liquid is the same as that of the molecules in the gas; at the same temperature, the liquid and gas phase differ mainly by the difference in the density between them. The magnitude of surface tension, γ, is determined by the internal forces in the liquid; thus, it will be related to the internal energy or cohesive energy.

It has been explained that the difference between the molecules at the surface and bulk arises from the number of near neighbors (Figure A.2). In other words, there should be a correlation between the heat of evaporation and surface tension. This was indeed found from a simple geometric model. The quantity ratio (i.e., surface

**TWO
DROPS**

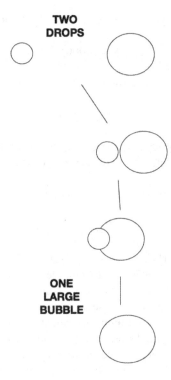

**ONE
LARGE
BUBBLE**

FIGURE 3.8   The smaller drop (or bubble) will merge into the larger drop due to the difference in the Laplace pressure.

tension and heat of evaporation) was found to be approximately ½ (as expected from
the Figure 3.2).

The phenomenon of surface tension can be explained by assuming that the surface behaves like a stretched membrane, with a force of tension acting in the surface
at right angles, which tends to pull the liquid surface away from this line in both
directions. Surface tension thus has units of force/length = mass distance/time 2 distance = mass/time². This gives surface tension in units as:

- mN/m,
- dyn/cm, or
- Joule/m² (mN m/m²).

For example (water): Volume per Mole in Gas or Liquid Phase and Distance between
Molecules in Gas and Liquid Phases

$$\text{Molar volume of water (at } 20°C) = 18 \text{ mL/mol} \qquad (3.3)$$

$$\text{Molar volume of water vapor} = 24.000 \text{ mL/mol} \qquad (3.4)$$

$$V_{gas} : V_{liquid} = ca \ 1000$$

Distance (D) between molecules in gas (D_G) or liquid (D_L) phase: (3.5)

$$D_{gas}: D_{liq} = (1000)^{1/3} = 10$$

This shows the quantity surface tension, γ, as the free energy excess per unit area:

$$\gamma = G_{surface}/area \qquad (3.6)$$

where $G_{surface}$ is the free energy of the two-phase system (phases a and b). The liquid and vapor phases are separated by a surface region (Adamson & Gast, 1997; Birdi, 2002, 2015, 2020).

It is also seen that other thermodynamic quantities would be given as:

$$\gamma = U_{surface} - S_{surface} \qquad (3.7)$$

Hence, the magnitude of surface tension, γ, is thus equal to the work needed in forming unit surface area (m^2 or cm^2). The magnitude of this quantity: work: increases the potential energy or surface free energy, G_s ($J/m^2 = erg/cm^2$), of the system. This can be further explained by different observations one makes in everyday life, where liquid drops contract to attain minimum surfaces. It is well known that the attraction between two portions of a fluid decreases very rapidly with the distance and may be taken as zero when this distance exceeds a limiting value, R_c, the so-called range of molecular action. Analyses have shown that (Rowlinson & Widom, 2003; Chattoraj & Birdi, 1984; Birdi, 2015, 2020) surface tension, γ, is a force acting tangentially to the interfacial area, which equals the integral of the difference between the external pressure, p_{ex}, and the tangential pressure, p_t:

$$\gamma = [integral] \, (p_{ex} \, p_t) \, dz \qquad (3.8)$$

the z-axis is normal to the plane interface and goes from the liquid to the gas. The magnitude of work that must be used to remove a unit area of a liquid film of thickness t will be proportional to the tensile strength (latent heat of evaporation) of the liquid thickness. In the case of water, this would give approximately 25,000 atm of pressure (600 cal/g = ca. 25.2×10^9 erg = 25,000 atm).

3.3 HEAT OF SURFACE FORMATION AND HEAT OF EVAPORATION OF LIQUIDS

In any physical phenomenon it is important to determine (knowledge) the heat of phase change.

The thermodynamics of surface tension of liquids requires the analyses of heat of surface chemistry formation. As mentioned earlier, energy is required to bring a molecule from the bulk phase to the surface phase of a liquid. In the bulk phase, the number of neighbors (six near-neighbors for hexagonal packing and if considering only two-dimensional packing) will be roughly twice the molecules at the surface (three near neighbors, when discounting the gas phase molecules) (See Figure 3.2).

The interaction forces between the surface molecules and the gas molecules will be negligible, since the distance between molecules in the two phases will be very large. Furthermore, these interaction differences disappear at the critical temperature. It was argued[7,] that when a molecule is brought to the surface of a liquid from the bulk phase (where each molecule is symmetrically situated with respect to each other), the work done against the attractive force near the surface will be expected to be related to the work spent when it escapes into the vapor phase. It can be shown that this is just half for the vaporization process (Figure 3.9).

Furthermore, there is a correlation between the latent heat of evaporation, L_{evap}, and γ or the specific cohesion, $a_{co}^2 \left(2\,\gamma/\rho_L = 2\,\gamma\,v_{sp} \right)$, where ρ_L = density of the fluid and v_{sp} is the specific volume. The following correlation was given:

$$L_{evap}\,(V_m)^{3/2}/a_{co}^2 = 3 \qquad (3.9)$$

However, later analyses showed that this correlation was not very satisfactory for experimental data. From these analyses it was estimated that there are 13,423,656 layers of molecules in $1\,cm^3$ of water (Adamson & Gast, 1997; Birdi, 2020) (at STP (e.g. standard temperature and pressure).

It is well known that both the heat of vaporization of a liquid, ΔH_{vap}, and the surface tension of the liquid, γ, are dependent on temperature and pressure, and they result from various inter-molecular forces existing within the molecules in the bulk liquid. In order to understand the molecular structure of liquid surfaces, one may consider this system in a somewhat simplified model.

The amount of heat required to convert $1\,g$ of a pure liquid into saturated vapor at any given temperature is called the latent **heat of evaporation** or **latent heat of vaporization**, L_{evap}. It has been suggested that:

$$\text{latent heat of evaporation}/2\gamma = L_{evap}/2\gamma \qquad (3.10)$$

From this model one can derive the following:

$$\text{Diameter} \times A_{mol} = 2\,\gamma/L_{evap} \qquad (3.11)$$

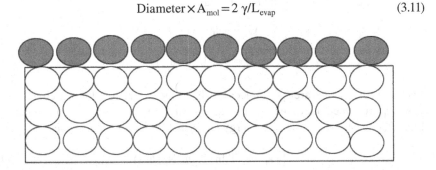

FIGURE 3.9   Molecular packing in two-dimensions in bulk and surface (shaded) molecules (schematic).

From this one finds

$$\text{Diameter of water molecule} = 2 \times 0.088 \times 1 / (2,541,300)$$
$$= 0.07 \text{ nm}$$
(3.12)

which is of the right order of magnitude.

In a later investigation, a correlation between heat of vaporization, ΔH_{vap}, and the effective radius of the molecule, R_{eff}, and surface tension, γ, was found. These analyses showed that a correlation between enthalpy and surface tension exists that is dependent on the size of the molecule. It thus confirms the molecular model of liquids.

3.4 SURFACE TENSION OF LIQUID MIXTURES

All industrial and natural liquid systems are made up of more than one component, which makes the studies of mixed liquid systems important. Further, the natural crude oil consists of a variety of alkanes (besides other organic molecules). The analyses of surface tension of liquid mixtures (for example, two or three or more components) have been the subject of studies in many reports.

Furthermore, it has been found that not all liquid mixture fluids exhibit ideal properties (Chattoraj & Birdi, 1984; Rowlinson & Widom, 2003).

According to one of these models of liquid surfaces, the free energy of the molecule is given as (Adamson & Gast, 1997; Birdi, 1997, 2002, 2015, 2020):

$$a_i = Ni \, g_i$$
(3.13)

where a_i is the absolute activity. This latter term can be expressed as

$$G_i = g_i \, s_i$$
(3.14)

where N_i is the mole fraction (unity for pure liquids) and g_i is derived from the partition function.

The free energy can thus be rewritten as:

$$G_i = g_i s_1$$
$$= k_B T \ln \left(a_1 / a_{1s} \right)$$
(3.15)

where s_1 is the surface area per molecule. This is the free energy for bringing the molecule, a_1, from the bulk to the surface, a_1. In a mixture consisting of two components, 1 and 2, one can derive the free energy terms as follows for each species:

$$\gamma \, s_1 = k_B \, T \ln (N_1 \, g_1 / N_1 \, s/g_1 s)$$
(3.16)

and

$$\gamma \, s_2 = k_B \, T \ln (N_2 \, g_2/N_2 \, s/g_2 s) \quad (3.17)$$

where N_s is the mole-fraction in the surface such that

$$N_1 \, s + N_2 \, s = 1 \quad (3.18)$$

It is safe to assume as a first approximation that: $s_m = s_1 = s_2$; that is, the surface area per molecule of each species is approximately the same. This will be reasonable to assume in such cases as mixtures of hexane + heptane, for example. This gives:

$$\gamma_s = k_B \, T \, (\ln (N_s \, g_1 \, g_{1s}) + \ln (N_s/g_{2s}) \quad (3.19)$$

Or, in combination with equation 3.22, one can rewrite as follows:

$$\exp(-\gamma_s/k_B T) = N_1 \exp(-\gamma_1 s/k_B T) + N_2^2 \exp(-g_2 s/k_B T) \quad (3.20)$$

Using the regular solution theory, the relation between activities was given as

$$R \, T \ln f_1 = -a_1 N_2^2; R \, T \ln f_2 = -a_1 N_1^2 \quad (3.21)$$

where f_1 denotes the activity coefficient.

In the case of some typical liquid mixtures, (useful models), a simple linear (ideal) relationship has been observed:

1. iso-octane – benzene mixtures: The surface tension changes gradually throughout. This means that the system behaves almost as an ideal. It is assumed that if the molecular packing geometry is similar, then the two liquids mix in the ideal state. Thus ideal mixtures indicate that both the geometric packing and intramolecular forces are interacting ideal conditions.
2. Water – electrolyte mixtures: The surface tension data of water-NaCl mixtures showed that the magnitude of γ **increases** linearly from ca. 72 to 80 mN/m for 0 – to 5-M NaCl solution (d γ/ mol NaCl = 1.6 mN/mol NaCl) (at STP: standard temperature and pressure).

The aqueous solutions of water – $NH_4 \, NO_3$ also showed an increase in γ, with the increase in concentration of NH_4NO_3.

It was found that the increase in γ per mol added NaCl, for aqueous solutions, is much larger (1.6 mN/m mol) than that for NH_4NO_3 (1.0 mN/m mol) (at STP). Similar differences are observed in the case of various electrolytes in aqueous solutions (Birdi, 2020).

In general, the magnitude of surface tension of water increases on the addition of electrolytes, with a very few exceptions. This indicates that the magnitude of surface excess term is different for different solutes. In other words, the state of solute molecules at the interface is dependent on the solute.

In the case of inorganic solutes in water, one observes some specific characteristics (Birdi, 2020).

The quantity: surface potential, S_p, is found to be related to the type of electrolyte added. The sign of Sp may change as follows (Birdi, 2020):

- zero value of $S_p > KCl + water$ solutions: This indicates that there are an equivalent number of anion – cation ions at the surface.
- positive value of S_p: This indicates that there is excess adsorption of cations than anions.
- negative value of S_p: These data indicate that there is an excess of cations than anions.

3.5 SOLUBILITY OF ORGANIC LIQUIDS IN WATER AND WATER IN ORGANIC LIQUIDS

The process of *solubility* of one compound into another is of fundamental importance in everyday life: examples are industrial applications (paper, oil, paint, and washing) and pollution control (oil spills, waste water control, toxicity, biological processes such as medicine, etc.). Accordingly, many reports are found in the literature that describes this process both on a theoretical basis and by using simple empirical considerations. The molecular picture of the system is very important for the understanding of the mechanisms. As already described here, the formation of a surface or interface requires energy; however, how theoretical analyses can be applied to curvatures of a *molecular-sized cavity* is not satisfactorily developed. It is easy to accept that any solubility process is in fact the procedure where a solute molecule is placed into the solvent where a cavity has to be made. Experiments show that these phenomena are not breaking any hydrogen-bonds in water. In other words, the heat of solution is endothermic.

The cavity has both a definite surface area and volume. The energetics of this process is thus a surface phenomenon, even if of molecular dimensions (i.e., nm²). The solubility of one compound, S, in a liquid such as water, W, means that molecules of S leave their neighbor molecules (SSS) and surround with WWW molecules. Thus, the solubility process means formation of a *cavity* in the water bulk phase where a molecule, S, is placed (WWWSWWW). It has been suggested that this cavity formation is a surface free energy process for solubility (Birdi, 2002, 2015).

The *solubility of various organic liquids in water* and vice versa is of much interest in different industrial and biological phenomena of everyday importance. In any of these applications, one would encounter instances where a prediction of solubility would be of interest (Tanford, 1980; Birdi, 2020).

Furthermore, solubilities of molecules in a fluid are determined by the free energy of solvation. In more complicated processes such as catalysis, the reaction rate is related to the de-solvation effects. A correlation between the solubility of a solute gas and the surface tension of the solvent liquid has been described, which was based on the curvature dependence of the surface tension for C_6H_6, C_6H_{12}, and CCl_4. This was based on the model that a solute must be placed in a hole (or cavity) in the solvent.

The change in the free energy of the system, ΔG_{sol}, transferring a molecule from the solvent phase to a gas phase is then:

$$\Delta G_{sol} = 4 \pi r^2 \gamma_q \, ej \tag{3.22}$$

where e_i is the molecular interaction energy. By applying the Boltzmann distribution law

$$c_{gs}/c_g = \exp(-\Delta G_G/k_B T) \tag{3.23}$$

where c_{gs} is the concentration of gas molecules in the solvent phase and c_g is their concentration in the gas phase. Combining these equations gives:

$$\ln\left(c_g^s/c_g\right) = \left(-4\pi r^2 \gamma_{aq}/k_B T\right) + e_i/k_B T \tag{3.24}$$

This model was tested for the solubility data of argon in various solvents, where a plot of log (Oswald coefficient) vs. surface tension was analyzed. In the literature, similar linear correlations were reported for other gas (e.g., He, Ne, Kr, Xe, O_2) solubility data.

The solubility of water in organic solvents does not follow any of these aforementioned models.

For instance, while the free energy of solubility, ΔG_{sol}, for alkanes in water is linearly dependent on the alkyl chain, there exists no such dependence of water solubility in alkanes.

3.6 THE HYDROPHOBIC EFFECT (IN AQUEOUS SOLUTIONS)

All natural processes are in general dependence on the physicochemical properties of water (especially when considering that over 70% of the Earth is covered by water: oceans are sometimes deeper than 5 km). Another important example is that all living species are both made and dependent on water.

Amphiphile molecules, such as long chain alcohols or acids, detergents, lipids, or proteins, exhibit *polar-apolar* characteristics, and the dual behavior is given this designation. The solubility characteristics in water are determined by the alkyl or apolar part of these amphiphiles, which arise from the hydrophobic effect. Some of the carbonaceous substances exhibit hydrophobic properties.

Hydrophobicity plays an important role in a wide variety of phenomena, such as solubility in water of organic molecules, oil – water partition equilibrium, detergents, washing, and all other cleaning processes, biological activity, drug delivery, and chromatography techniques. Almost all drugs are designed with a particular hydrophobicity as determined by the partitioning of the drug in the aqueous phase and the cell lipid membrane.

The ability to predict the effects of even simple structural modifications on the aqueous solubility of an organic molecule could be of great value in the development of new molecules in various fields, for example, medical or industrial. There exist theoretical procedures to predict solubilities of nonpolar molecules in nonpolar solvents and for salts or other highly polar solutes in polar solvents, such as water or similar substances. However, the prediction of solubility of a nonpolar solute in water has been found to require some different molecular considerations (Birdi, 1982; Tanford, 1980).

Furthermore, the central problems of living matter comprise the following factors: recognition of molecules leading to attraction or repulsion, fluctuations in the force of association and in the conformation leading to active or inactive states, the influence of electromagnetic or gravitational fields and solvents including ions, and electron or proton scavengers. In the case of life processes on Earth, one is mainly interested in solubility in aqueous media.

The unusual thermodynamic properties of nonpolar solutes in the aqueous phase were analyzed, by assuming that water molecules exhibit a special ordering around the solute. This water-ordered structure was called the *iceberg structure*. The solubility of semi-polar and non-polar solutes in water has been related to the term molecular surface area of the solute and some interfacial tension term.

The solubility, X_{solute}, in water was derived as:

$$R\,T\,(\ln X_{solute}) = -\,(\text{surface area of solute})\,(\gamma_{sol}) \qquad (3.25)$$

where surface tension, γ_{sol}, is some *micro – interfacial tension* term at the solute-water (solvent) interface (Figure 3.10).

The quantity surface area of a molecule is the cavity dimension of the solute when placed in the water medium.

The data of solubility, total surface area (TSA), and hydrocarbon surface area (HYSA) have been analyzed for some typical alkanes and alcohols. The relationship between different surface areas of contact between the solute solubility (sol) and water was derived as:

$$\ln\,(sol) = -0.043\,\text{TSA} + 11.78/(\text{RT})\Delta G_{o,sol} = -\text{RT}\,\ln\,(sol) = 25.5\,\text{TSA} + 11.78 \qquad (3.26)$$

where *sol* is the molar solubility and TSA is in Å^2.

The quantity 0.043 (RT = 25.5) is some (arbitrary) micro-surface tension. It is also important to mention that at the molecular level there cannot exist any surface property that can be uniform in magnitude in all directions. Hence, the micro-surface tension will be some average value.

In the case of alcohols, assuming a constant contribution from the hydroxyl group, the magnitude of the quantity:

hydrocarbon surface area (HYSA) = TSA – hydroxyl group surface area:

$$\ln\,(sol) = -0.0396\,\text{HYSA} + 8.94 \qquad (3.27)$$

However, one can also derive a relationship that includes both HYSA and OHSA (hydroxyl group surface area):

$$\ln\,(sol) = -0.043\,(\text{HYSA})\,(0.06\,\text{OHSA}) + 12.41 \qquad (3.28)$$

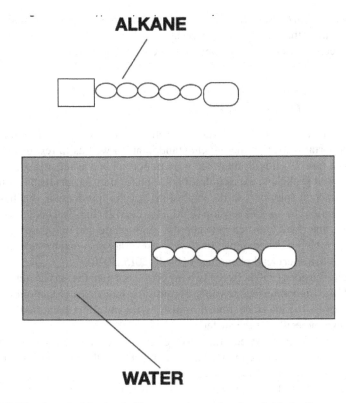

FIGURE 3.10 A typical hydrophobic non polar molecule soluble in the aqueous phase. Cavity formation.

The relations described above did not give correlations to the measured data that were satisfactory (ca. 0.4 to 0.978). The following relationship was derived based on the solubility data of both alkanes and alcohols, which gave correlations in the order of 0.99:

$$\ln (sol) = 0.043 \ HYSA + 8.003 \ IOH - 0.0586 \ OHSA + 4.42 \qquad (3.29)$$

where the IOH term equals 1 (or the number of hydroxyl groups) if the compound is an alcohol and zero if the hydroxyl group is not present.

The term HYSA thus can be assumed to represent the quantity that relates to the effect of the hydrocarbon part on the solubility. The effect is negative, and the magnitude of t is 17.7 erg/cm^2.

The magnitude of OHSA is found to be 59.2 Å^2. As an example, the surface areas of each carbon atom and the hydroxyl group in the molecule 1-nonanol were estimated.

$$CH_3 \quad CH_2 \quad CH_2 \quad CH_2 \quad CH_2 \quad CH_2 \quad CH_2 \quad CH_2 \quad OH$$

84.9/31.8/31.8/31.8/31.8/31.8/31.8/31.8/OH

It is seen that the surface area of the terminal *methyl* group (84.9 Å^2) is approximately three times larger than the methylene groups (31.82 Å^2, or 31.82 10^{-20} m^2). Computer simulation techniques have been applied to such solution systems.

3.7   INTERFACIAL TENSION OF LIQUIDS (LIQUID$_1$ – LIQUID$_2$; OIL – WATER)

The interfacial forces present between two phases, such as immiscible liquids, are of much importance from a theoretical standpoint, as well as in regard to practical systems. The liquid$_1$ – liquid$_2$ interface is an important one as regards such phenomena as chemical problems, extraction kinetics, phase transfer, emulsions (oil-water), fog, and surfactant solutions. In the case of primary oil production, one has to take into consideration the surface tension of oil. However, during a secondary or tertiary recovery, the interfacial tension between the water phase and oil phase becomes an important parameter. For example, the *bypass* and other phenomena such as snap-off are related to the interfacial phenomena (Birdi, 2015, 2020).

Interfacial tension (IFT) between two liquids is less than the surface tension of the liquid with the higher surface tension, because the molecules of each liquid attract each other across the interface, thus diminishing the inward pull exerted by that liquid on its own molecules at the surface.

The precise relation between the surface tensions of the two liquids separately against theory vapor and the interfacial tension between the two liquids depends on the chemical constitution and orientation of the molecules at the surfaces. In many cases, a rule proposed by Antonows holds true with considerable success (Birdi, 2002, 2020).

3.8   LIQUID – LIQUID SYSTEMS: WORK OF ADHESION

The surface tension is the force that is present between two different phases (Adamson & Gast, 1997; Birdi, 2002, 2015). The free energy of interaction between dissimilar phases is the work of adhesion, W_A (energy per unit area):

$$W_A = W_{AD} + W_{AP} \qquad (3.30)$$

where W_A is expressed as the sum of different intermolecular forces, for example,

   a. London dispersion forces, D_L;
   b. Hydrogen bonds, H_b;
   c. Dipole-dipole interactions, D_D;
   d. Dipole-induced interactions, D_I;
   e. Π bonds, Π;
   f. Donor-acceptor bonds, D_A;
   g. Electrostatic interactions, E_L.

It is also found that the W_{AD} term will always be present in all systems (i.e., liquids and solids), while the other contributions will be present to a varying degree as determined by the magnitude and nature of the dipole associated with the molecules. In order to simplify the terms given by the above equation, one procedure has been to compile all the intermolecular forces arising from the dipolar nature of W_{AP}.

The calculated value of surface tension of n-octane was analyzed from these parameters.

The calculated value for γ of n-octane $= 19.0$ mN/m, while the measured value is 21.5 mN/m, at 20°C (i.e., $\gamma_{octane} = \gamma_{LD}$). The real outcome of this example is that such theoretical analyses do indeed predict the surface dispersion forces, γ_{LD}, as measured experimentally, to a good accuracy. In a further analysis, the Hamaker constant, A_i, for liquid alkanes is found to be related to γ_{LD} as:

$$A_i = 3 \times 10^{-14} \, (\gamma_{LD})^{11/12} \tag{3.31}$$

This was further expanded to include components at an interface between phases I and II:

$$A_{I,II} = 3 \times 10^{-14}/e_2 \left(\gamma_I^D - \gamma_{II}^D \right)^{11/6} \tag{3.32}$$

where e_2 is the dielectric constant of phase 2; however, in some cases, forces other than dispersion forces would also be present. The manifestation of intermolecular forces is a direct measure of any interface property and requires a general picture of the different forces responsible for bond formation, as discussed in the following.

a. Ionic bonds: The force of attraction between two ions is given as:

$$F_{ion} = (g^+ \, g^-)/r^2 \tag{3.33}$$

and the energy, U_{ion}, between two ions is related to r_{ion} by the equation

$$U_{ion} = (g + g^-)/r_{ion} \tag{3.34}$$

where two charges (g^+, g^-) are situated at a distance of r_{ion}.

b. Hydrogen bonds: Based on the molecular structure, those conditions under which hydrogen bonds might be formed are (a) presence of a highly electronegative atom, such as O, Cl, F, and N, or a strongly electronegative group such as -CCl$_3$ or -CN, with a hydrogen atom attached; (b) in the case of water, the electrons in two unshared sp^3 orbitals are able to form hydrogen bonds; (c) two molecules such as $CHCl_3$ and acetone (CH_3COCH_3) may form hydrogen bonds when mixed with each other, which is of much importance in interfacial phenomena.

c. Weak-electron sharing bonding: In magnitude, this is of the same value as the hydrogen bond. It is also the Lewis acid-Lewis base bond (comparable to Brønsted acids and bases). Such forces might contribute appreciably to cohesiveness at interfaces; a typical example is the weak association of iodine (I_2) with benzene or any polyaromatic compound.

The interaction is the donation of the electrons of I_2 to the electron-deficient aromatic molecules (π-electrons).

d. Dipole-induced dipole forces: In a symmetrical molecule, such as CCl_4 or N_2, there is no dipole (ma = 0) through the overlapping of electron clouds from another molecule with dipole, m_b, with which it can interact with induction. It will thus be clear that various kinds of interactions would have to be taken into consideration whenever we discuss interfacial tensions of liquid – liquid or liquid – solid systems (Adamson & Gast, 1997; Kwok et al., 1994).

3.9   INTERFACIAL TENSION THEORIES OF LIQUID – LIQUID SYSTEMS

As shown above, various types of molecules exhibit different intermolecular forces, and their different force and potential-energy functions can be estimated (Birdi, 1997, 2002). If the potential-energy function were known for all the atoms or molecules in a system, as well as the spatial distribution of all atoms, it could in principle then be possible to add up all the forces acting across an interface.

Further, this would allow one to estimate the adhesion or wetting character of interfaces. Because of certain limitations in the force field and potential-energy functions this is not quite so easily attained in practice. Further, the direct microscopic structure at a molecular level is not currently known. However, various indirect molecular data are available from different techniques: light refraction; contact angle at a liquid-solid interface (Chapter 3 4).

For example, to calculate the magnitude of surface tension of a liquid, one needs knowledge of the radial pair-distribution function. However, for the complex molecule, this would be highly difficult to measure, although data for simple liquids such as argon have been found to give the desired result. The intermolecular force in saturated alkanes arises only from London dispersion forces. Now, at the interface, the hydrocarbon molecules are subjected to forces from the bulk molecule, equal to γ. Also, the hydrocarbon molecules are under the influence of London forces due to molecules in the oil phase. It has been suggested that the most plausible model is the geometric means of the force due to the dispersion attraction, which should predict the magnitude of the interaction between any dissimilar phases.

As described earlier, the molecular interactions arise from different kinds of forces, which means that the measured surface tension, γ, arises from a sum of dispersion, γ_D, and other polar forces, γ_P (Chapter 3):

$$\gamma = \gamma_D + \gamma_P \tag{3.35}$$

Here, γ_D denotes the surface tensional force solely determined by the dispersion interactions, and γ_P arises from the different kinds of polar interactions (Equation 3.35). The interfacial tension between hydrocarbon (HC) and water (W) can be written as:

$$\gamma_{HC,W} = \gamma_{HC} + \gamma_W - 2\,(\gamma_{HC}\,\gamma_{W,D})^{1/2} \tag{3.36}$$

where subscripts HC and W denote the hydrocarbon and water phases, respectively. Considering the solubility parameter analysis of mixed-liquid systems, it is found that the geometric mean of the attraction forces gives the most useful prediction values of interfacial tension. Analogous to that analysis in the bulk phase, the geometric mean should also be preferred for the estimation of intermolecular forces at interfaces. The geometric mean term must be multiplied by a factor of two since the interface experiences this amount of force by each phase. However, the relation in Equation 3.432 was alternatively proposed by Antonow (Birdi, 1997, 2002, 2020):

$$\gamma_{12} = \gamma_1 + \gamma_2 - 2 \, (\gamma_1 \, \gamma_2)^{1/2} = ((\gamma_1)^{1/2} - (\gamma_2)^{1/2})^2 \qquad (3.37)$$

This relation is found to be only an approximate value for such systems as fluorocarbon – or hydrocarbon – water interfaces, while not applicable to polar organic liquid – water interfaces.

In order to analyze these systems, a modified theory was proposed (Chapter 3). The expression for interfacial tension was given as (Adamson & Gast, 1997; Birdi, 1997):

$$\gamma_{12} = \gamma_1 + \gamma_2 - 2 \, \Phi \, (\gamma_1 \, \gamma_2)^{1/2} \qquad (3.38)$$

where the value of Φ varied between 0.5 and 0.15. Φ is a correction term for the disparity between molar volumes of v_1 and v_2. This theory was extensively analyzed in the literature, and a satisfactory agreement was found with experimental data.

3.10 ANALYSIS OF THE MAGNITUDE OF THE DISPERSION FORCES IN WATER (γ_D)

One finds various kinds of liquids in nature. This is related to the liquid physical properties on planet earth.

Water is known to play a very important role in a variety of systems encountered in everyday life, and its physicochemical properties are of much interest. Therefore, the magnitude of water γ_D has been the subject of much investigation and analysis. By using Equation 3.35, and the measured data of interfacial tension for alkanes-water, the magnitude of γ_D has generally been accepted to be 21.8 mN/m (at 25°C).

In order to obtain any thermodynamic information of such systems it is useful to consider the effect of temperature on IFT. The alkane-water IFT data have been analyzed.

These data show that IFT is lower for $C_6 H_{14}$ (50.7 mN/m) than for the other higher-chain-length alkanes. The slopes (interfacial entropy: $-d\gamma/dT$) are all almost the same, ca. 0.09 mN/m per CH_2 group. This means that water dominates the temperature effect, or that the surface entropy of IFT is determined predominantly by the water molecules. Further, as described earlier, the variation of surface tension of alkanes varies with chain length. This characteristic is not present in IFT data; however, it is worth noting that the slopes in IFT data are lower than those of both pure alkanes and water.

3.11 LIQUID – SOLID SYSTEMS (CONTACT ANGLE – WETTING – ADHESION) UNDER DYNAMIC CONDITIONS

The state of liquid in contact with a solid surface is of much importance in many everyday phenomena (detergency, adhesion, friction, wetting, flotation, suspensions, solid emulsions, erosion, printing, pharmaceutical products, oil/gas reservoirs, etc.). If one considers two systems, such as a drop of liquid (water) placed on different solid surfaces (glass, Teflon), one observes the following. The contact angle, θ, (Figure 3.11), is defined by the balance between surface forces (surface tensions) between the respective phases, solid (γ_S), liquid (γ_{liquid}), and LS (liquid – solid) (γ_{SL}) (Chapter 3) (Birdi et al., 1989; Birdi & Vu, 1993; Birdi, 2020):

In the case of water (drop) – glass and water (drop) – Teflon, one finds that the magnitude of θ is 30° and 105°, respectively. Since the liquid is the same, then the difference in contact angle arises from the different solid surface tensions. From this one can therefore conclude that the surface tension of a solid is an important surface parameter (Chapter 3). Further, the relation between Young's equilibrium contact angle and the **hysteresis** on rough paraffin wax surfaces was investigated.

Advancing and receding contact angles (contact angle hysteresis) of four organic liquids and water were measured on a variety of polymer surfaces and silicon wafers using an inclinable plane. The magnitude of contact angles varied widely from liquid to liquid and from surface to surface. Surface roughness was relatively unimportant. Instead, the contact angles seemed to be more closely tied to the chemical nature of the surfaces. In general, contact angles increased with the liquid surface tension and decreased with the surface tension of the solid. Several definitions were used to calculate contact angle hysteresis from the experimental data. Although hysteresis is usually considered an extensive

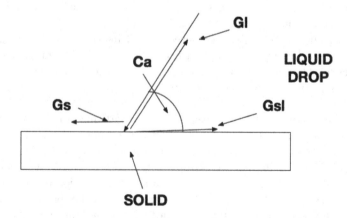

Contact Angle (CA)
Equilibrium of
Surface Tension Forces

FIGURE 3.11   Contact angle (θ = CA) at the liquid – solid interface (surface tension of liquid = Gl; surface tension of solid = Gs; surface tension of solid-liquid = Gsl).

property, it was found that on a given surface a wide range of liquids gave a unique value of reduced hysteresis. Apparently, reduced hysteresis represents an intrinsic parameter describing liquid – solid interactions.

In some studies, the **dynamic systems of liquid drops,** when placed on a smooth solid surface, have been investigated. The system liquid drop – solid is a very important system in everyday life. For example, this is significant in rain drops placed on tree leaves or other surfaces. It is also significant in all kinds of systems where a spray of fluid is involved, such as in sprays or combustion engines. Further, in many diagnostics devices (such as blood analyses), enzymes are applied in very small amounts (microliters) from solutions as a drop on an electrode. This means that after evaporation of the fluid the remaining enzyme must be well described for the diagnostic instrument to function reliably. The dynamics of a liquid drop evaporation rate is of much interest in many phenomena (combustion engines, rain drops and environment, aerosols, and pollution). The liquid – solid interface can be considered as follows. Real solid surfaces are, of course, made up of molecules not essentially different in their nature from the molecules of the fluid. The interaction between a molecule of the fluid and a molecule of the boundary wall can be regarded as follows. The molecules in the solid state are not as mobile as those of the fluid. It is therefore permissible for most purposes to regard the molecules in the solid state as stationary. However, complexity arises in those liquid – solid systems where a layer of fluid might be adsorbed on the solid surface, such as in the case of water – glass.

Systematic studies have been reported in literature on the various modes of liquid drop evaporation when placed on smooth solid surfaces (Birdi, 2002, 2015). In these studies, the rate of the mass and contact diameter of water and n-octane drops placed on glass and Teflon surfaces were investigated. It was found that the evaporation occurred with a constant spherical cap geometry of the liquid drop. The experimental data supporting this were obtained by direct measurements of the variation of the mass of droplets with time and by the observation of contact angles. A model based on the diffusion of vapor across the boundary of a spherical drop was considered to explain the data. Further studies were reported, where the contact angle of the system was $\theta < 90°$. In these systems, the evaporation rates were found to be linear and the contact radius constant. In the latter case, with $\theta > 90°$, the evaporation rate was non-linear, the contact radius decreased, and the contact angle remained constant.

As a model system, one may consider the evaporation rates of fluid drops placed on polymer surfaces in still air. The mass and evaporating liquid (methyl acetoacetate) drops on the polytetrafluoroethylene (Teflon) surface in still air have been reported. These studies suggested two pure modes of evaporation: at a constant contact angle with a diminishing contact area and at a constant contact area with a diminishing contact angle. In this mixed mode the drop shape would vary, resulting in an increase in the contact angle with a decrease in the contact circle diameter, or, sometimes a decrease in both quantities.

The data for the state of a liquid drop placed on a smooth solid surface can be described in terms of the

- adius,
- height of the drop,
- weight,
- and the contact angle, θ.

The liquid drop when placed on a smooth solid can have a spherical cap shape, which will be the case in all cases where the drop volume is approximately 10 μL or less. In the case of much larger liquid drops, ellipsoidal shapes may be present, and different geometrical analyses will have to be implemented. In the case of rough surfaces, one may have much difficulty in explaining the dynamic results. However, one may also expect that there will be instances where it is non-spherical. This parameter will need be determined before any analyses can be carried. In the case of a liquid drop that is sufficiently small, surface tension dominates over gravity. The liquid drop can then be assumed to form a spherical cap shape. A spherical cap shape can be characterized by four different parameters, the drop height (h_d), the contact radius (r_b), the radius of the sphere forming the spherical cap (R_s), and the contact angle (θ).

The data showed that only in some special cases the magnitude of θ remains constant under evaporation, such was in a water – teflon or water-glass system (Birdi, 2020) (Figure 3.12).

The liquid film that remains after most of the liquid has evaporated has been investigated (Birdi, 2020). It was shown that one could estimate the degree of porosity of solid surfaces from these data. This method just shows that one can determine the porosity of a solid in a very simple manner as compared to other methods. These data show that one can determine the porosity of solids, without the use of a mercury porosimeter or other method. The latter studies are much more accurate, since they were based on measurements of change of weight of drop vs. time under evaporation.

3.12   SURFACE CHEMISTRY CHARACTERISTICS OF SOLID SURFACES

3.12.1   INTRODUCTION

It is useful to delineate a short description of evolutionary surface chemistry of the planet earth in the following. This is essential since some terminology is not used extensively in the current literature as regards the system: climate change/sun-atmosphere -earth system/carbon recycling.

It is known that based on the geological time-scale (about few billions of year ago!) the planet earth came in existence as a very hot ball during the evolution. Accordingly, this was the pre-life era (Calvin, 1969). This simple evolutionary observation also convincingly indicates that the planet earth has cooled (and most likely still changing (such as: earthquake eruptions; etc)) down to its present temperature limits:

- crust temperatures range: −50°C (Himalaya) to +50°C (Sahara)
- interior temperature: over 5,000°C (lava-like)

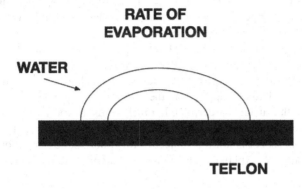

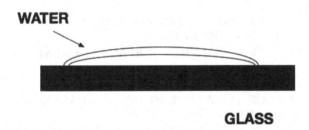

FIGURE 3.12   Profiles of liquid drops under evaporation (snap shots at two time intervals): water on Teflon; water on glass.

This is also observed in studies where climate temperature is reported to be changing *(dynamic).

This stage of evolution of earth was different from what one experiences today. This indicates that surroundings of earth have been changing continuously since its creationism. Furthermore, one finds that some climate aspects are still changing: both due to natural and man-made interactions.

EXAMPLE

>EARTH EVOLUTION STAGES (INTRODUCTION)
       <0>STAGE CREATION (VERY HIGH TEMPERATURES) (PRE WATER-PLANTS)
   <1>STAGE PRE LIVING SPECIES
       <2>STAGE BEGIN LIVING SPECIES

This also indicates that everything surrounding the planet earth is changing (non-linear) since its creation. In other words, climate (among other phenomena) is continuously changing (depending on time/place). Further, mankind has no measuring yardstick to analyze these geological aged stages of evolution.

At sometimes, during its evolution, the surface of the earth started to cool and the crust (consisting of land/oceans/rivers/lakes) was formed. The cooling era started at the surface and formed a crust, while the interior is still known to remain liquid-like and very hot (lava). This is evident from the fact that earthquakes release this lava in certain areas on the surface of earth (). In other words, it would be very difficult for mankind to assess whether these changes are currently going on and to what extent.

Although, this aspect is not the subject of this book, but a very short mention is necessary to add some explanation about the evolutionary status stages.

EXAMPLE

SIMPLIFIED STAGES OF THE EARTH STATES AS FOUND TODAY
>CREATION OF EARTH: LAVA-LIKE (FLUID) HOT PLANET: ALMOST SPHERICAL
    >PRE-WATER/ PRE-LIVING SPECIES/PRE-PLANTS
    .>VERY HIGH TEMPERATURE
>>COOLING   OF   EARTH   SURFACE: CRUST   FORMATION:
PLAINS-HILLS-MOUNTAINS
    >SURFACE CHEMISTRY OF SOLID
>>>APPEARANCE OF WATER IN OCEANS/LAKES/RIVERS: ICE AT NORTH/SOUTH POLES
    >>>APPEARANCEOFATMOSPHEREWITHLOWCONCENTRATION
OF CARBON DIOXIDE: APPEARANCE OF PLANTS/INSECTS/ANIMALS/
OTHER LIVING SPECIES/MANKIND
>>>>CURRENT STAGE IS THUS CHANGING

The phenomenon of cooling and crust formation is complex. It is suggested to be related to the effect of atmosphere on the magnitude of sun heat. In addition, earthquake eruptions further suggest that heat (lava) moves from the interior of the earth to its surface and cools to for crust.

The geological description of such a vast system is obviously out of scope of this manuscript.

The crust of earth is thus, a combination of a solid (land) and liquid (i.e. water: oceans/rivers/lakes), and the ice-covered polar regions.

Furthermore, the surface of earth is covered by 75% oceans, and the latter is found to vary in depth (ranging up to over 5 km depth).

Most of the large oceans are found in the Southern Hemisphere. Furthermore, it is found that due to continuously lava eruptions around the world, the movement of fluid-like lava from the interior of earth to the surface is not completely stopped, which indicates that different phenomena: such as climate is still changing.

EXAMPLE

EARTH (SURFACE – INTERIOR)
...
    EARTH   SURFACE -> LAND.......FORESTS........RIVERS.......LAKES....... OCEANS.....
    ANTARCTIC**/NORTH-POLE.....
    EARTH INTERIOR -> SOLID------FLUID (HOT-LAVA)
....

FURTHER, THE FLUID PROPERTY OF OCEANS IS REFLECTED IN THE TIDAL EFFECTS AS one observes from the moon movement (attraction) around the earth. These phenomena are nonlinear and any future modeling would only be an estimate.

3.12.2 SURFACE TENSION OF SOLIDS

In any process where two different phases (in the present case: solid – gas) are involved, the surface chemistry of the substances becomes very important. In every-day life one observes distinctly, the difference between Teflon and other solids, as regards the wetting or adhesion characteristics. Further, in complex structures, such as coal, the surface properties are found to depend on the composition (Chattoraj & Birdi, 1984; Birdi, 2020).

Solid surfaces exhibit some specific characteristics which are of much different properties from those of the liquid surfaces (Adamson & Gast, 1997; Chattoraj & Birdi 1984; Birdi, 2004, 2016, 2020).

In all processes where solids are involved, the primary process is dependent on the surface property of the solid. In some distinct ways the solid surfaces are found to be different than the liquid surfaces (Chattoraj & Birdi, 1984; Birdi, 2020). One finds a large variety of applications where the surface of a solid plays an important role (for example: active charcoal, talc, cement, sand, catalysis, oil, and gas reservoirs (shale reservoirs), plastics, wood, glass, clothes and garments, biology (hair, skin, etc), road surfaces, polished surfaces, friction (drilling technology), etc). Solids are rigid structures and resist any stress effects. It is thus seen that many such consideration here in the case of solid surfaces will be somewhat different than for liquids. Further, in all porous solids, the flow of gas or liquid oil, means that interfaces (liquid – solid) are involved. Hence, in such a system the molecular interactions of surfaces chemistry of the phases are of importance. Many important technical and natural (such as: earthquakes) processes in everyday life are dependent on the rocks (etc) in the interior of earth. This thus requires the understanding of the surface forces on solid interfaces (solid – gas; solid – liquid ; solid$_1$ – solid$_2$).

The surface chemistry of solids has been described based on the classical theories of chemistry and physics (Adamson & Gast, 1997; Birdi, 2002, 2016; Birdi, 2020). Another example is the corrosion of metals (interaction of oxygen in air with certain metals: iron/aluminum/copper/zinc/etc) which initiates at surfaces, thus requiring treatments which are based upon surface properties. As described in the case of liquid surfaces, analogous analyses of solid surfaces have been carried out. The molecules at the solid surfaces are not under the same force field as in the bulk phase, Figure 3.13.

In the bulk phase, each solid atom is surrounded symmetrically by near neighboring atoms. However, at the surface, toward the air (for example), there is asymmetric situation. This asymmetry imparts special properties of solids at their surfaces.

Experiments show that the solid surface is the most important characteristic. The differences between perfect ones and surfaces with defects are very obvious in many everyday observations. For example, the shine of all solid surfaces increases as the surface becomes smooth. Further, the friction decreases between two solid surfaces as the solid surfaces become smooth. The solids were the first materials which were analyzed at a molecular scale (by using X-ray diffraction, etc). This leads to the understanding of the structures of solid substances and the crystal atomic structure. This is because while molecular structures of solids can be investigated by such methods as X-ray diffraction, the same analyses for liquids are not that straightforward. These analyses have also shown that there exist surface defects at the molecular level.

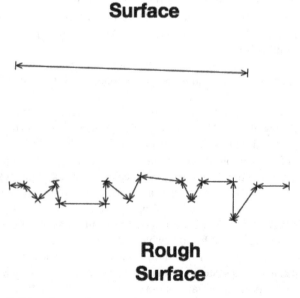

Smooth Surface

Rough Surface

FIGURE 3.13 Solid surface molecular defects: (a) perfect crystal (smooth); (b) surface with defects (rough).

As pointed out for liquids (Section 3.2), one will also consider that when the surface area of a solid powder is increased by grinding (or some other means), then surface energy (energy is supplied to the system) is needed. Of-course, due to the energy differences between solid and liquid phases, these processes will be many orders of magnitude different from each other. The liquid state of course retains some structure which is similar to its solid state, but in the liquid state the molecules exchange places. The average distance between molecules in the liquid state is roughly 10% larger than that in its solid state (Chapter 1). It is thus desirable at this stage to consider some of the basic properties of liquid solid interfaces. The surface tension of a liquid becomes important when it comes in contact with a solid surface. The interfacial forces which are present between a liquid and solid can be estimated by studying the shape of a drop of liquid placed on any smooth solid surface (Figure 3.2 addd). The shape (or the angle) of the drop of a particular liquid on different solid surfaces is found to be different. In other words, this has been useful in determining the solid surface tension.

3.12.3   Solid Surface Forces (Wetting Properties of Solid Surfaces)

In various surface chemistry phenomena where solid surfaces are involved, the characteristics of a solid surface are of importance. There is no direct procedure to estimate the solid surface property, as one finds for liquids (i.e. measurement of surface tension) (Chapter 3.4).

As an example, one may observe the difference between solid surface tensions, for example glass or TEFLON, which is seen in the following example (Figure 3.14).

It is seen that a drop of water shows a very low contact angle, while on the surface of TEFLON, it has a magnitude of 105°.

Another phenomenon of interest is the **degree of wetting of a solid surface.** The degree of **wetting** when a liquid comes in contact with a solid surface is the most common phenomenon in everyday life (washing and detergency; water flow in underground; rain water seepage; cleaning systems; water flow in rocks; oil/gas recovery

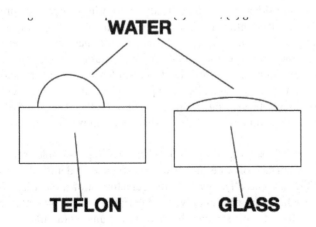

FIGURE 3.14   A drop of water on: (a) Teflon; (b) glass.

in shale; etc). The liquid-solid and gas-solid interactions are of highly significant characteristics in everyday life; e.g. gas/oil reservoirs (Birdi, 2020).

The liquid and solid surface interface can be described by considering a classical example. Wetting of solid surfaces is well known when considering the difference between Teflon and metal surfaces. To understand the degree of wetting, between the liquid, L, and the solid, S, it is convenient to rewrite the Young's equation as follows:

$$Cos\ (\theta) = (\gamma_S - \gamma_{LS})/\gamma_L \qquad (3.40)$$

which would then allow one to understand the variation of γ_S with the change in the other terms. The latter is important because complete wetting occurs when there is no finite contact angle, and thus $\gamma_L <> \gamma_S - \gamma_{LS}$. However, when $\gamma_L > \gamma_S - \gamma_{LS}$, then Cos $(\theta) < 1$, and a finite contact angle is present. The latter is the case when water, for instance, is placed on hydrophobic solid, such as Teflon, polyethylene, or paraffin. The addition of surfactants to water, of course, reduces γ_L, therefore, 0 will decrease on the introduction of such surface active substances (Adamson & Gast, 1997; Chattoraj and Birdi, 1984; Birdi, 2002, 2016). The state of a fluid drop under dynamic conditions, such as evaporation, becomes more complicated (Birdi and Vu, 1989; Birdi, 2020). However, in the present it is useful to consider the spreading behavior of when a drop of one liquid is placed on the surface of another liquid, especially when the two liquids are immiscible.

The spreading phenomenon was analyzed by introducing a quantity, spreading coefficient, $S_{a/b}$, defined as Adamson & Gast, 1997; Birdi, 2002, 2016, 2020):

$$S_{a/b} = \gamma_a\ (\gamma_b + \gamma_{ab}) \qquad (3.41)$$

where $S_{a/b}$ is the spreading coefficient for liquid b on liquid a, γ_a and γ_b are the respective surface tensions, and γ_{ab} is the interfacial tension between the two liquids. If the value of $S_{a/b}$ is positive, spreading will take place spontaneously, while if it is negative, liquid b will rest as a lens on liquid a.

However, the value of γ_{ab} needs to be considered as the equilibrium value, and therefore if one considers the system at non-equilibrium, then the spreading coefficients would be different. For instance, the instantaneous spreading of benzene is observed to give a value of $S_{a/b}$ as 8.9 dyn/cm, and therefore benzene spreads on water. On the other hand, as the water becomes saturated with time, the value of (water) decreases and benzene drops tend to form lenses. The short chain hydrocarbons such as hexane and hexene also have positive initial spreading coefficients and spread to give thicker films. Longer chain alkanes, on the other hand, do not spread on water, e.g. the $S_{a/b}$ value for C16H34 (n-hexadecane)/water is 1.3 dyn/cm at 25°C (Table 3.1).

The spreading of a solid (polar organic) substance, e.g. example cetyl alcohol (C18 H_{38} OH), on the surface of water has been investigated in detail (Adamson & Gast, 1997; Birdi, 2020). Generally, however, the detachment of molecules of the amphiphile into the surface film occurs only at the periphery of the crystal in contact with the air water surface. In this system, the diffusion of an amphiphile through the bulk water phase is expected to be negligible, because the energy barrier now includes not

TABLE 3.1

Calculation of Spreading Coefficients, $S_{a/b}$, for Air Water Interfaces [20°C] [a = air; w = water; o = oil]

Oil	$\gamma_{w/a}$	$\gamma_{o/a}$	$\gamma_{o/w} = S_{a/b}$	Conclusion
n-$C_{16}H_{34}$	72.8	30.0	52.1 = 0.3	will not spread
n-Octane	72.8	21.8	50.8 = +0.2	will just spread
n-Octanol	72.8	27.5	8.5 = +36.8	will spread

only the formation of a *hole* in the solid, but also the immersion of the hydrocarbon chain in the water. It is also obvious that the diffusion through the bulk liquid is a rather slow process. Furthermore, the value of $S_{a/b}$ would be very sensitive to such impurities as regards the spreading of one liquid upon another.

Another example is that the addition of surfactants (detergents: surface active agents) to a fluid dramatically affects its wetting and spreading properties (Chapter 3). Thus, many technologies utilize surfactants for control of wetting properties (Birdi, 1997, 2020). The ability of surfactant molecules to control wetting arises from their self-assembly at the liquid-vapor, liquid-liquid, solid-liquid, and solid-air interfaces and the resulting changes in the interfacial energies (Birdi, 1997). These interfacial self-assemblies exhibit structural details and variation. The molecular structure of the self-assemblies and the effects of these structures on wetting or other phenomena remain topics of extensive scientific and technological interest.

As an example, in the case of *oil spills* on the seas, these considerations become very important. The treatment of such pollutant systems requires the knowledge of the state of the oil. The thickness of the oil layer will be dependent on the spreading characteristics. The effect on ecological phenomena (such as: birds; plants) will depend on the spreading characteristics.

Contact angle at liquid₁-sold-liquid₂ interface:
In addition, there are also many systems where one finds:

$$LIQUID_1 - SOLID - LIQUID_2$$

One typical example is: oil (liquid₁) spills on the oceans (liquid₂), where one has:

$$oil - water - solid.$$

The *Youngs* equation at liquid1 – solid – liquid₂ has been investigated for various systems. This is found in such systems where the liquid₁ – solid – liquid₂ surface tensions meet at a given contact angle. For example, the contact angle of water drop on Teflon is 50° in octane (Chattoraj & Birdi, 1984; Adamson & Gast, 1997; Birdi, 2003, 2016, 2020) (Figure 3.15):

......water.....Teflon........octane

**Water - Octane - Teflon
Contact Angle (CA)**

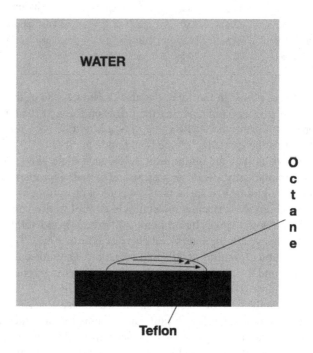

FIGURE 3.15   Contact angle at water – Teflon – octane interface.

In this system the contact angle, θ, is related to the different surface tensions as follows:

$$\gamma_{\text{s-octane}} = \gamma_{\text{water-s}} + \gamma_{\text{octane-water}} \, \text{Cos} \, (\theta) \tag{3.42}$$

or:

$$\text{Cos} \, (\theta) = (\gamma_{\text{s-octane}} \, \gamma_{\text{water-s}})/\gamma_{\text{octane-water}} \tag{3.43}$$

This gives the value of θ = 50°, when using the measured values of ($\gamma_{\text{s-octane}} \, \gamma_{\text{water-s}}$)/ $\gamma_{\text{octane-water}}$. The experimental value of θ (= 50°) is the same. This analysis showed that the assumptions made in derivation of Young's equation are reasonably correct.

4 Carbon Capture – Recycling and Storage (CCRS) (Basic Remarks)

4.1 INTRODUCTION

In this chapter, the parameters related to man-made effects on any *carbon dioxide change in its equilibrium concentration* (especially in atmosphere) and climate (increasing temperature (only surface temperature) are explained. The perturbation by any man-made activities is considered in this aspect. The *chemical evolutionary equilibrium* (**CEE**) on earth has stabilized over geological ages. The current man-made perturbation to this CEE is being debated (IPCC, 2011; Filho, 2022; Gates, 2022; Birdi, 2020; Rosenzweig et al., 2021), as regards surface temperature change.

The parameter temperature of earth is complex and therefore the time and place need to be defined. In the current text, the subject deals with observations related to analyses of temperature and climate as related to the surface of earth. Although, it should be mentioned that both the earth and the universe are dynamic, especially, as regards climatic conditions, over geological periods of time.

After the surface temperature on earth had stabilized, such that photosynthesis could be observed. This means that CEE had achieved conditions, which were stable for the growth of plants. The addition of man-made carbon dioxide, CO_{2gas}, has caused perturbation to the CEE. Although, CO_{2gas} is known to be increasing at the present state. This is also known to have an effect on the current GHG, related to carbon dioxide in atmosphere. However, the phenomenon of GHG, related to carbon dioxide, CO_{2gas}, has been known to be in existence since photosynthesis was observed. The latter phenomenon has been known to have existed for about a billion years on earth. Therefore, any perturbation to CEE by mankind is of consideration.

The man-made fossil fuel combustion, since the advent of industrial revolution, has given rise to following:

- equilibrium concentration of CO_{2gas} in atmosphere (420 ppm: 0.042%)
- most combustion processes produce flue gases (10,000 ppm: 10%)

These concentrations are comparatively low and need capture treatment, before it can be treated. In other words, carbon capture is the most necessary primary phenomenon.

Current methods are being used/developed:

EXAMPLE

<RECYCLING
<adsorption on solids
<absorption in liquids
<Storage in suitable spent reservoirs
<PARTIAL RECYCLING
<PLASTICS (Household/ Cars/ Packaging/Construction)
<Foods/Fisheries--metabolism by man

These capturing methods give rise to production of almost >90% carbon dioxide. After a suitable capturing process, the gas is treated in different means.

In the case of recycling of captured carbon dioxide, there are various processes.

In the case of partial recycling, there are already some large industrial applications. For example, in the car industry, almost 30%–40% of the body parts consist of plastics (i.e. carbon dioxide). If a car average life is over 10–15 years, then the partial recycling is already in progress. There are currently about 3 billion cars worldwide. However, if this plastic material was recycled, after this period, then almost all is achievable.

EXAMPLE

<PARTIAL CARBON RECYCLING IN CAR BODY PARTS (WORLDWIDE)

Number of cars/trucks (worldwide) $= 3 \ 10^9$
Amount of plastic (= carbon) each vehicle $= 100\,kg$
Total plastic/carbon $= 3 \ 10^9 \times 100$ kg $= 3 \ 10^9 \times 0.1$ ton
$= 3 \times 10^8$ ton
$= 300$ million tons

In the case of storage, which is currently the largest method, the gas is injected in an appropriate spent reservoir. The captured gas, CO_{2gas}, is generally compressed at low temperature, where it turns as $CO_{2liquid}$. The latter, which is generally in liquid form, reduces the volume by about a factor of 1,000 (Figure 4.1).

EXAMPLE

<MAN-MADE FOSSIL FUEL COMBUSTION AND CARBON DIOXIDE (CO_{2gas}) EMISSION CYCLE

<FOSSIL FUELS COMBUSTION ONLY TAKES PLACE ON LAND AREAS

<DIFFUSION OF CARBON DIOXIDE FROM LAND AREAS TO SURFACE OF OCEANS

<ABSORPTION OF CARBON DIOXIDE IN OCEANS/LAKES/ RIVERS

CAPTURE OF CARBON DIOXIDE

l

FROM ATMOSPHERE (420 ppm)

l

FROM FLUE GAS (10%)

l

l

ADSORPTION
OR
ABSORPTION

FIGURE 4.1   Carbon (CO_{2gas}) capture by different methods: adsorption or absorption + recycling + other methods.

The current concentration of carbon dioxide in the earth environment is considered to be high. This has resulted from the increasing combustion process of fossil fuels over the past 200 years, since the beginning of the industrial revolution. In other words, the concentration of CO_{2gas} in the atmosphere has been changing since it was observed in the atmosphere (almost a billion years ago). However, there is very little data about the concentrations of different GHG gases.

Thus, the capture techniques are considered.

EXAMPLE

<SYSTEM – S U N@ATMOSPHERE@ E A R T H>

In this system, heat is input from the billions of year source: planet sun. It is reported that latter quantity, however of input, is nonlinear and variable.

EXAMPLE

< DYNAMIC SUN HEAT RADIATION>

<SUN: SPOTS – SUN ROTATION AND PATH VARIABLE

The next stage of interaction arises from the sun – atmosphere *interface*.

However, this interface is varying, as regards the state of atmosphere (i.e. weather conditions).

<SUN>> ATMOSPHERE (CLEAR/CLOUDY/RAIN/POLLUTION)>>
EARTH SURFACE (LAND (30%) (DESERTS/FORESTS?/ANTARCTIC) –
OCEANS+LAKES+RIVERS (70%)

The heat balance which mainly originates from sun has to interact through these different phases (interfaces). The importance of application of surface and colloid chemistry becomes obvious and also essential in CCRS.

The evolution of the solar system is of interest in this subject. The age of earth has been considered by many. The geological evolution of the solar system has been going on for over a period of a few billions of years. This geological evolution has main interactions between living species growth and existence on the earth (only). The interactions between one of the living species, mankind, have particularly bearing the climate effects on earth (Birdi, 2020). It is also mentioned that any man-made activity on earth, especially increasing usage of fossil fuels and technology, also adds pollution aspects which additionally affect the climate (Figure 4.2).

Over the billions of years, the carbon (carbonate) system has reached an equilibrium (pseudo) with its surroundings (i.e. air (i.e. CO_2 in air) – water – carbonate salts in the oceans) (Birdi, 2020). This means, as also is evident for other eco-systems on earth, the latter equilibrium is very large and does not fluctuate appreciably within short time periods. This observation is also evident from the CO_{2gas} data over many thousands of years. These data indeed show a very stable content with slow fluctuations. This means that the atmosphere (especially its composition) is a complex system. The composition of air near the surface of earth has been monitored for many decades. The concentration of nitrogen (78%) and oxygen (21%) is found to have remained fairly constant over many decades. However, one has analyzed and monitored the *evolutionary chemical equilibrium*, CO_2 concentration (varying from 280 ppm to the current 400 ppm) variation over a century. This has been carried out since it is known to be a GHG gas. Based on chemical equilibrium between one particular gas, CO_2, this has been of added interest at the present state in the literature.

EXAMPLE

MATERIAL BALANCE OF CARBON DIOXIDE ($CO_{2,GAS}$) ON EARTH

CO_2 (GAS)

IN ATMOSPHERE (CO_{2gas}) (750 Gton-GHG)
IN OCEANS/LAKES/RIVERS/CO_{2aq}) (39000 Gton-GHG)

IN PLANTS (PHOTOSYNTHESIS) (X Gton-GHG) (on earth and oceans/lakes/rivers)
IN FLUE GASES (MAN-MADE) (25 Gton-GHG)
IN INTERIOR OF EARTH (unknown Gton-GHG)

............(the Gton magnitudes are only estimates: Gton = 1 billion tons)

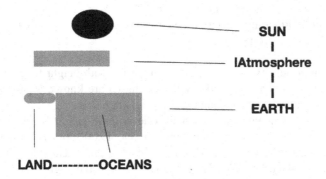

FIGURE 4.2 Different environmental Stages: Pre- and Current: Carbon dioxide (CO_2) equilibria: air – land areas (plants) – oceans/lakes (carbonate).

The analyses are about the change in composition of GHG gases in the atmosphere. For instance, the current equilibrium concentration of CO_{2gas} (which is in equilibrium with concentration of CO_2 in oceans/lakes/rivers and other carbonates + plants) is ca.

400 ppm (0.04%). This quantity is an estimate with reference to a number of measuring places on earth (Birdi, 2020). These places are all near the surface of earth (Birdi, 2020). In other words, it is assumed that any change (increase/decrease) in the concentration of CO_{2gas} at these places is linearly changing at all heights from the surface of earth. This is not a complete analysis, as it omits a very poor assumption and disregards the nonlinearity in the system.

In addition, due to photosynthesis (whereby carbon dioxide from the atmosphere is captured by plants (on land) and in oceans) (Birdi, 2020).

However, studies assert that by year 2100 this estimated value in atmosphere (420 ppm) may increase to 1,000 ppm (IPCC, 2011; Dubey et al., 2002). This is argued from the fact that the current fossil fuel combustion adds (ca. 5 ppm/year) CO_{2gas} to the atmosphere. This is also based on the assumption that:

- current fossil fuel amount remains almost the same (which depends on availability and degree of control (i.e. changing to non-fossil energy sources: such as: atomic energy; hydro-energy; wind-energy; solar-energy). These different energy sources are not equivalent in some subtle characteristics.

Therefore, more detailed and extensive studies are found in the literature, which are related to capture and control of CO_{2gas} from air or during combustion of fossil fuels at the energy plants (flue gases), in order to mitigate the CO_{2gas} concentration (increase) phenomena. It is important to mention that there are studies which show that the concentration of CO_{2gas} has some significant characteristics (Birdi, 2020):

- **150 ppm**: plants may stop to grow.
- **280 ppm**: pre-industrial average (minimum CO_{2gas} concentration needed for plant growth and to sustain life on earth).
- **410 ppm**: current equilibrium value (CEE).
- **1,000–2,000 ppm**: variation some thousands year ago (absence of man-made combustion).

The molecular density, in atmosphere, gets lower with height from the surfaces of earth (Chapter 1). Thus trees (i.e. photosynthesis) are known to stop growing at a given height in hills. This is found around 2–4 km heights.

Further, it is known that carbon dioxide is toxic gas at higher concentrations (Appendix A). The toxicity for mankind is ca. 19.000 ppm (1.9%). Mankind has existed on earth for over hundred thousand years. This means that the equilibrium concentration of carbon dioxide has never been in the higher toxic range. From this one may conclude that CEE has induced the criterion that in atmosphere was originally.

The technology of carbon capture and storage (**CCRS**) (or carbon capture and sequestration or carbon control and sequestration (Fanchi & Fanchi, 2016; IPCC, 2011; Birdi, 2020) is described as the process of capturing waste carbon dioxide (CO_2) from large point sources, such as fossil fuel power plants, transporting it to a storage site, and depositing it where it will not enter the atmosphere, normally an underground geological formation. The aim is to prevent the release of large quantities of CO_{2gas} (man-made) into the atmosphere (from fossil fuel use in power generation and other industries). It is a potential means of mitigating the contribution of fossil fuel emissions to global warming. Generally, estimates show:

- power plants deliver 500–1,000 MW of electric power.
- It is estimated that a 1,000 MW (MW = million W) plant produces: Coal: ca. 7 Mt/year of CO_2; Oil: 5 Mt/year of CO_2; Gas: 3 Mt/year of CO_2.
- Wind/Solar Energy sources thus (intermittently) produce 500–1,000 MW with very little carbon dioxide emissions: there is though need for production/maintenance/backup-energy-source.

Furthermore, it is known that gaseous CO_{2gas} is *soluble* in water (i.e. CO_{2aq}) (as a weak acid). In addition, CO_{2aq} interacts with water to form $H_2 CO_3$. All these substances (carbonates) do not exhibit GHG characteristics.

This means that CO_{2gas} is soluble in the oceans (lakes; rivers). Earth is covered by oceans more than 70% of the earth's surface. Oceans thus are found to contain a very large amount of CO_2. Furthermore, CO_{2gas} dissolves in water and (interacts) forms bi-carbonate (HCO_3^-) and carbonate (CO_3) ions. Since oceans (with depths of several kilometers)/lakes comprise vast quantities of water, it is obvious that there will be large amounts of CO_2/bi-carbonate/carbonate. Carbon dioxide (CO_2) as present in oceans (lakes), however, does not contribute to the GHG phenomena. In addition, the carbonate cycle in oceans/lakes/rivers further contributes to both inorganic (shells; rocks) and biological systems (fisheries).

4.2   CARBON DIOXIDE AQUEOUS SOLUBILITY EQUILIBRIA IN (WATER) OCEANS/LAKES (WATER)

It is important to mention that some substances, especially carbon dioxide (CO_2) exhibit unique physico-chemical properties (Appendix A).

EXAMPLE

UNIQUE CHEMICAL PROPERTIES OF CARBON DIOXIDE (CO_2)
- Lethal effects on living creatures above 1% concentration of CO_{2gas} in air
- Essential (minimum suggested lime) concentration (approximately 200–500 ppm) ideal to provide plants/food on planet earth
- Mankind (other living species) have existed for long time – this shows that the equilibrium concentration of carbon dioxide has remained under the lethal plateau.
- Minimum concentration in air, about 200 ppm, for photosynthesis to grow plants/food/fisheries/shells: during evolutionary era.
- Geological studies show:

- photosynthesis and GHG phenomena initiated pre-living species; i.e. billions of years ago
- currently photosynthesis and GHG and living species are known to be at CEE.

The carbon dioxide (CO_2) solubility equilibria in oceans/lakes have been reported in literature (Rackley, 2010; Birdi, 2020). A brief mention is made, here, in order to explain the CO_2 concentration in atmosphere under the evolution.

The following equilibria are present:

>CARBON DIOXIDE SOLUBILITY EQUILIBRIUM:

CO_{2gas} ---- $CO_{2aqueous}$:
    Carbon dioxide (gas: in air) / / in equilibrium / / Carbon dioxide (gas: in water)

$$CO_{2gas} = CO_{2aq} \qquad (4.1)$$

EXAMPLE

CARBON DIOXIDE MAGNITUDES IN DIFFERENT PHASES AT EQUILIBRIUM:
    It is estimated that the magnitudes of carbon dioxide in different phases are very different:

PHASE	AMOUNT
$CO_{2,air}$	750 Gtons
$CO_{2,oceans}$	40,000 Gtons

(estimated quantities) (pseudo-equilibrium state in oceans/lakes/rivers)

The equilibrium constant is defined as:

$$K_{CO_2} = \left[CO_{2aq} \right] / \left[x_{CO_2} \right] \left[CO_{2gas} \right] \left[p_{CO_2} \right] \qquad (4.2)$$

where: K_{CO_2} is the equilibrium constant; $CO_{2aqueous}$ is the concentration of CO_2 in water (mol/kg); x_{CO_2} is the mole fraction of CO_2 (in air); CO_{2gas} is the concentration in air; p_{CO_2} is the partial pressure of carbon dioxide.

$CO_{2aqueous}$ and water interact (instantly):

$$CO_{2aqueous} + H_2O = H^+ + HCO_3^- \qquad (4.3)$$

With an equilibrium constant:

$$K_{HCO_3} = [H^+][HCO_3^-]/[CO_{2aq}] \tag{4.4}$$

And the dissociation equilibrium constant:

$$K_{CO_3} = [H^+][CO_3--]/[HCO_3^-] \tag{4.5}$$

Where: K_{CHCO_3} is the equilibrium constant; (H^+) is the concentration of hydrogen ion; x_{CO_2} is the mole fraction; CO_{2gas} is the concentration in air; p_{CO_2} is the pressure.

These equilibria show that during the evolution, concentrations of CO_2 in air and oceans have reached an equilibrium. The man-made CO_2 (i.e. fossil fuel combustion) is very small as compared to the carbon stored in the oceans/lakes. However, the carbonate/bicarbonate equilibria are therefore capturing parts of man-made CO_2 from air. The total carbon content in oceans is thus the sum of all the different quantities:

$$[CO_{2aqueous}] + [CO_3^{-2}] + [HCO_3^-] \tag{4.6}$$

It is also found from these equilibria that any change in the concentration of CO_2 in air would result in a change in the components in these equilibria. The carbonates in oceans thus stabilize the average concentration of CO_2 in air. These carbonates are known to be a buffer property.

The photosynthesis process (which captures CO_2) is known to be much slower than the equilibria in oceans. However, the most significant observation from these data is as follows:

a. CO_{2gas} concentration in air is controlled mainly by the equilibria in oceans. This arises from the fact that the total inorganic carbon in the oceans (e.g. $CO_{2aqueous} + HCO_3 + CO_3$: with ratios of approximately: 1:10:100) is a few decades times larger than CO_{2gas} in air.

b. CO_{2gas} concentration in air is thus always comparatively **low** due to these equilibria. This may be due to some evolutionary processes which have brought about these constraints on concentration of CO_2. Therefore, data also indicate that the concentrations of CO_{2gas} have always been lower than 10,000 ppm (1%) over the past many thousands of years (i.e. below toxicity levels).

c. Photosynthesis and forests (trees, plants, etc) capture CO_{2gas} from air, and the equilibria under (a) keep the minimum CO_{2gas} in air. This also indicates that GHG phenome has existed.

d. Solubility equilibrium of CO_{2gas} in air and rain drops (CO_{2aq}.)

e. All living species use plants/food/fisheries and exhale CO_{2gas}.

These data are merely mentioned here, since there have been reported some studies which investigated the possibility of carbon storage in deep oceans (which is stratified). Analyses of CO_2 in oceans show that due to lack of mixing (at very deep structures), this is a pseudo-equilibrium (Schulz et al., 2006; Birdi, 2020).

Most importantly, the CEE phenomenon thus predicts that:

- CEE has induced global chemical CO_{2gas} equilibrium in the environment.
- the equilibrium constant, K_{CO_2} (equation 4.2), maintains the CEE.
- any change in CO_{2gas} will thus perturb the state of equilibrium.

After CO_2 is recovered from flue gas (or directly from air), it needs to be stored or used in some recycled application (such as: food/drinks/enhanced oil recovery (EOR); etc). As regards the CO_2 storage, it has been injected into geological formations for several decades for various purposes, including enhanced oil recovery. CO_2 has also been stored in suitable old reservoirs.

Storage of the captured CO_2 is currently also being envisaged either in deep geological formations, or in the form of mineral carbonates (as: M_xCO_3). It is also possible technically to store CO_2 in deep oceans. However, the CO_2 solubility will lead to acidification. It is also known that due to large deep structures of oceans, there is not complete mixing of CO_2 (as absorbed from air). Intact, one finds large stratifications and the oceans are only at psuedo-equilibria with CO_2 in air. This parameter has many implications. The atmosphere – ocean interface constitutes a very fast equilibrium. This arises from the fact that CO_{2gas} molecules dissolve in water and interact with H_2O:

$$[CO_{2gas}] == [CO_{2aq}] \tag{4.7}$$

$$CO_{2aq} + H_2O == H_2CO_3 \tag{4.8}$$

In addition, it is expected that *carbon capture recycling storage* technology (CCRS) applied to a modern conventional power plant could reduce CO_2 emissions to the atmosphere by approximately 80%–90% compared to a plant without CCS. It is (IPCC, 2005) estimated that the economic potential of CCS/CCRS could be between 10% and 55% of the total carbon mitigation application until year 2100. However, these climate models have ignored the observation about the changes in parameters (Kemp et al., 2022).

Carbon dioxide can be captured out of air or fossil fuel power plant flue gas using adsorption (or carbon scrubbing), membrane gas separation, or other adsorption technologies (such as: cryogenic process; mineral capture; solution storage in deep oceans; etc) (Keith et al., 2018).

Absorption of CO_2 by amines is currently the leading carbon scrubbing technology. Capturing and compressing CO_2 may increase the energy needs of a coal-fired CCRS plant by 25%–40%. The carbon capture is obviously a costly addition to the industrial sector where fossil fuels are burned (combustion process). These and other system costs are estimated to increase the cost per watt energy produced by 21%–91% for fossil fuel power plants.

It has been suggested that, with the development of relevant research, development, and deployment, sequestered coal-based electricity generation in 2025 may cost less than un-sequestered coal-based electricity generation today. Geological formations are currently considered the most promising sequestration sites.

A general problem is that long term predictions about submarine or underground storage security are very difficult and uncertain, and there is still the risk that CO_2 might leak into the atmosphere (Birdi, 2020). Accordingly, it is therefore necessary to understand the adsorption mechanisms.

However, the technology of capturing CO_{2gas} is reported to be most effective at point sources, such as large fossil fuel or biomass energy facilities, industries with major CO_2 emissions, natural gas processing, synthetic fuel plants, and fossil fuel-based hydrogen production plants. Extracting CO_2 from air is also a viable process.

Flue gas from the combustion of coal in oxygen (from air) has a relatively large concentration of CO_2, about 10%–15% CO_{2gas} whereas natural gas power plant flue gas is about 5%–10% CO_2 (McDonald et al., 2015).

Therefore, it is more energy and cost efficient to capture CO_2 from coal-fired power plants. Impurities in CO_2 streams, like sulfur and water, could have a significant effect on their phase behavior. In instances where CO_2 impurities exist, especially with air capture, a scrubbing separation process would be needed to initially clean the flue gas. The captured CO_{2gas} is liquified (under low temperatures/high pressure), thus reducing its volume/weight by about 1000x times.

4.3 FOSSIL FUEL COMBUSTION PROCESS

Since combustion of fossil fuels produces CO_2, it is obvious that this process has been investigated in recent reports. For example, it has been suggested that, by gasifying coal, it is possible to capture approximately 65% of carbon dioxide embedded in it and sequester it in a solid form. The **combustion** process of fossil fuels is given as:

$$\text{Fossil fuel} + \text{oxygen (from air)} == CO_{2gas} + H_2O \\ + \text{diverse (other) pollutant gases} \tag{4.9}$$

The diverse pollutants gases, which are produced, are NO_x, CO, and SO_2. In general, one finds three different configurations of technologies for carbon capture:

- post-combustion, pre-combustion, and oxy-fuel combustion.
  - In post combustion capture, CO_2 is removed after combustion of the fossil fuel – this is the scheme that would be applied to fossil-fuel burning power plants. Here, carbon dioxide is captured from flue gases at power stations or other large point sources. The technology is currently at an advanced stage and is used in other industrial applications, although not at the same scale as might be required in a commercial scale power station. Post combustion capture is most popular in research because existing fossil fuel power plants can be retrofitted to include CCS technology in this configuration (Sumida et al., 2012).
  - The technology for pre-combustion is widely applied in fertilizers, chemicals, gaseous fuels (H_2, CH_4), and power production. In these cases, the fossil fuel is partially oxidized, for instance in a gasifier. The resulting syngas (CO and H_2) is shifted into CO_2 and hydrogen (H_2). The resulting CO_2 can be captured from a relatively pure exhaust stream.

Hence, hydrogen (H_2) can now be used as fuel; the carbon dioxide is removed before combustion takes place. There are several advantages and disadvantages when compared to conventional post combustion carbon dioxide capture (Birdi, 2020), integrated gasification combined cycle for carbon capture storage.

- Carbon dioxide (CO_{2gas}) is captured and removed after combustion of fossil fuels. This scheme is applied to new fossil fuel burning power plants, or to existing plants where re-powering is an option. The capture before expansion, i.e. from pressurized gas, is a normal process in almost all industrial CO_{2gas} capture processes, at the same scale as will be required for utility power plants.

4.4 CARBON CAPTURE AND RECYCLING AND STORAGE (CCRS) METHODS

4.4.1 CCRS TECHNOLOGY AND DIFFERENT METHODS OF CAPTURE

Besides the main methods of carbon capture: e.g. adsorption on solids and absorption in fluids (Chapters 2 and 4), there are various other procedures which one finds in the current literature. These will be delineated here briefly. Carbon recycling from fossil fuel combustion presses requires different technologies pertinent to each process.

EXAMPLE

>CARBON RECYCLING AND CAPTURE PROCESSES
>CARBON CAPTURE FROM ATMOSPHERE
>CARBON CAPTURE FROM FLUE GASES

It is well accepted that due to the urgent need to capture/recycling/capture/control of the carbon in atmosphere, there is a need to use some of these different techniques (Gates, 2021).

4.4.2 DIRECT CAPTURE OF CARBON (CO_{2GAS}) FROM AIR

In spite of different process, the current consensus is that the amount of carbon (CO_2) as gas is high, as compared to concentrations before the industrial revolution (Birdi, 2020).

Direct air capture from air is a process of removing CO_2 directly from the ambient air (as opposed to from point sources). Combining direct air capture with carbon storage could be useful as a carbon dioxide removal technology and as such would constitute a form of climate engineering if deployed at a large scale.

Although it is out of scope of this book, only a short mention is given as regards the cost of such a CCRS process is estimated to about two thirds of the total cost of CCS, making it limit the wide-scale deployment of CCRS technologies.

Another method which is reported is called the chemical looping combustion (CLC). Chemical looping uses a metal oxide as a solid oxygen carrier. Metal oxide particles react with a solid, liquid, or gaseous fuel in a fluidized bed:

$$\text{AIR } (CO_{2gas} \text{ } 0.04 \text{ }\%)\text{-----------CAPTURE TECHNOLOGY--------PURE } CO_2$$
$$(4.10)$$

Accordingly, some technological methods have been investigated for such direct air capture (Keith et al., 2018) methods.

Furthermore, an economic estimate of this carbon recovery application (in 2018) estimated the cost at ca. 150 USD per ton (1,000 kg) of atmospheric CO_2 captured. It is expected that such estimates will decrease as better innovation techniques are developed.

In these studies, the following are the most significant methods:

a. using alkali and alkali-earth hydroxides,
b. carbonate ions.
c. organic–inorganic hybrid sorbents consisting of amines supported in porous adsorbents.

The equilibrium concentration of CO_{2gas} in the atmosphere is highly diluted compared to point-source CO_2 capture. Accordingly, the capture costs will be expected to be high. On the other hand, the carbon recovery costs will decrease as the importance of its effect on climate change increases.

In the current literature, one finds that different forms have been conceived for permanent storage of CO_2. These forms include gaseous storage in various deep geological formations (including saline formations and exhausted gas fields), and solid storage by reaction of CO_2 with metal oxides to produce stable carbonates.

Geological storage of CO_2: This technology is about handling of captured CO_2. The captured CO_2 has been stored under varying procedures. This procedure has also been called as *geo-sequestration*; this method involves injecting carbon dioxide, generally in supercritical form, directly into underground geological formations. Oil fields, gas fields, saline formations, unmineable coal seams, and saline-filled basalt formations have been suggested as storage sites (Birdi, 2017, 2020). In a specific example, carbon dioxide capture is advanced by injecting the gas into spent oil reservoirs in the North Sea (Danish government) 2023.

4.4.3   Captured CO₂---------Storage (Oil Field/Saline Formations/ Unmineable Coal Seams)

Various physical, e.g., highly impermeable caprock and geochemical trapping mechanisms would prevent CO_{2gas} from escaping to the surface.

4.4.4   CO$_2$ Injection in Oil Reservoirs------Enhanced Oil Recovery (EOR)

In the oil recovery processes, carbon dioxide ($CO_{2gas/liquid}$) injection has been used for decades.

$CO_{2liquid}$ has been used in oil recovery processes (EOR). This process is sometimes used where CO_2 liquid is injected into declining oil fields to increase oil recovery. Approximately 200 million tons of $CO_{2liquid}$ are injected annually in the declined oil fields. This option is attractive because the geology of hydrocarbon reservoirs is generally well understood and storage costs may be partly offset by the sale of additional oil that is recovered. Oil recovery from reservoirs is a multistep process. Most of the recovery is primarily based on original pressure in the reservoir (producing around 20% of oil in place). Enhanced oil recovery (EOR) is a process which is used to increase the amount of crude oil that can be extracted from an oil field (Birdi, 2015). In carbon capture and sequestration enhanced oil recovery (CCS EOR), carbon dioxide is injected into an oil field to recover oil that is often never recovered using more traditional methods.

4.5   ADDITIONAL DIFFERENT CARBON CAPTURE METHODS

There are found additional approaches to determine the feasible application of technologies which can be viable for capturing and recycling/storage in the current literature (Herzog & Golomb, 2004; Birdi, 2020).

4.5.1   Carbon Dioxide Gas-Hydrate Recycling/Capture

The structures of water and ice are found to be abnormal, as compared to other liquid – solid systems. For instance, ice is about 10% lighter than water (that is why icebergs float on water). The simple reason for this characteristic is that the molecular structure of ice is such that the volume per water molecule is larger than that in water (at 0°C). This open structure of ice (as compared to the water structure at the same temperature: near the freezing point of water) is found to capture some gas molecules. These ice – gas structures are called gas-hydrates (Tanford, 1980; Birdi, 2017, 2020).

In nature, one finds that under suitable pressure and temperature, gas hydrates are present (in very large quantities). Most of these are methane hydrates. It is found that along continental margins worldwide, are located at and above the gas hydrate stability zones. It is thus likely that a few, if any, of these methane discharge events along continental slopes are perhaps related to anthropogenic climate change and global warming.

4.5.2   Other Carbon Capture Methods

Carbon Capture by using Algae (photosynthesis): It is found that a large number of industries have been using micro-algae produce viable foods (Herzog & Godrom, 2004; Birdi, 2020). This is found to amount to over 5,000 tons of food per year. It is also possible to convert the algae to bio-fuels.

Carbon capture using Ocean Fertilization: It has been suggested that by enhancing the fertilization in oceans (by using iron), the growth of marine phytoplankton.

This will be expected to partially sink to the bottom of oceans/lakes, and will remain there for many decades.

Carbon Capture by Mineral (e.g. carbonates) Storage: The physical properties of carbon dioxide (gas) indicate that it is a reactive gas, based on appropriate conditions. For example, many housing paints are based on the criteria that on exposure to carbon dioxide in air, the suitable components in the housing paints will react. This reaction generally forms a suitable carbonate salt (making the paint more resistant to weather).

Furthermore, it is known that many minerals as found in oceans are made up of carbonate (M_xCO_3), which uses (captures) carbon dioxide. Some examples are calcium magnesium silicates. The reaction with magnesium silicate (serpentine) is as follows:

$$Mg_3SiO_2(OH)_4 + 3\ CO_{2gas} = 3\ MgCO_3 + 2\ SiO_2 + 2\ H_2O \qquad (4.11)$$

Carbon dioxide (CO_{2gas}) Sources and Sinks: Carbon dioxide (CO_{2gas}) concentration in air is currently 400 ppm, and in other natural and man-made processes. It is appropriate to consider briefly the different CO_2 sources and sink parameters. Approximately, the combustion of fossil fuels worldwide is reported to produce an 30 Gt of CO_2 per year. Deforestation of tropical regions accounts for an additional 4 Gt CO_{2gas} per year. With the CO_2 natural (carbon) content cycle and also the related terrestrial and ocean water CO_{2aq} sinks (due to solubility in water), the annual increase in CO_2 emissions is reported to be approximately 15 Gt CO_2 per year. This quantity is almost equivalent to 2 ppm per year (concentrations of CO_2 in air). Fossil fuel-based emissions of CO_2 may be sourced from both stationary (e.g., power plant) and non-stationary systems (e.g., automobile (transport); etc). The amount emitted is approximately 13 Gt CO_2 per year on average from large stationary sources globally. In addition to CO_2 emission generation from the oxidation of fossil fuels, flue gases may also be sourced as a result of a chemical process. Although these emission sources represent a minor portion of total anthropogenic emissions, the chemical processing method currently used may be required for the formation of a useful product, such as cement or steel. Therefore, due to the difficulty of replacing CO_2-generating chemical processes with others that are absent of CO_2, it is crucial that these emission sources are not disregarded. The majority of fossil-fuel oxidation (combustion) processes produce the CO_2 emissions, while there is a fraction of emissions generated from chemical processes. Some typical examples are:

- cement manufacturing; iron and steel industries; oil and gas reservoirs; gas processing, oil refining, and ethylene production. Mitigation associated with the capture of CO_2 from these industrial-based processes is small compared to that of the transportation and electricity sectors; however, there may not be alternatives to the materials created from these processes, such as cement, iron, and steel production, etc.

In the following, some of these aspects are delineated (briefly):

- Cement manufacturing results in CO_{2gas} emissions sourced from *calcination* in addition to the fuel combustion emissions of the cement kilns.

  It is estimated that the emissions worldwide from cement industry are approximately 2–4 Gt of CO_2 with approximately 52% and 48% associated with the calcination process and cement kilns, respectively. Ordinary cement is a mixture of primarily di- and tri-calcium silicates ($2CaO \cdot SiO^2$, $3CaO \cdot SiO_2$) as well as small amounts of other compounds:

  [calcium sulfate ($CaSO_4$); magnesium, aluminum, and iron oxides (Fe_2O_3); and tri-calcium aluminate ($3CaO \cdot Al_2O_3$)].

  The primary reaction that takes place in this process is the formation of calcium oxide and CO_2 from calcium carbonate, which is highly endothermic and requires 3.5–6.0 GJ per ton of cement produced.
- The *steel-making indu*stry is a multi-step technology. This industry produces a combination of emissions and the chemical processes comprise the estimated 1 Gt of CO_2 emitted worldwide. Steel-making, generates CO_2 as a result of carbon oxidation to carbon monoxide, which is required for the reduction of hematite ore (Fe_2O_3) to molten iron (pig iron). The amount of CO_2 produced is much lower than CO_2.

  Another man-made source of the CO_{2gas} is also due to the combination of coal-burning and limestone calcination. In the second stage of the steel-making process, the carbon content of pig iron is reduced in an oxygen-fired furnace from approximately 4%–5% down to 0.1%–1%, and is known as the basic oxygen steelmaking (BOS) process. Both of these steps produce a steel-slag waste high in lime and iron content.
- In an oil refinery plant, crude oil, a mixture of various hydrocarbon components ranging broadly in molecular weight is fractionated from lighter to heavier components. In a second stage, the heavier components are catalytically "cracked" to form shorter hydrocarbon chains. In addition to producing CO_{2gas} as a byproduct of the distillation and catalytic cracking processes, the heat and electricity required for the methane reforming process used in H2 production for hydro cracking and plant utilities produce additional CO_{2gas}.
- Recovered natural gas from gas fields or other geologic sources often contains varying levels of non-hydrocarbon components such as CO_{2gas}, N_2, H_2S, and helium.

  Natural gas (primarily methane and ethane) and other light hydrocarbons such as propane and butane, to less extent, are the valuable products in these cases, and often generated CO_{2gas} is a near-pure stream. Concentrations of approximately 20% CO_2 are not uncommon in large natural gas fields.
- **Exhaust Emissions:** Comparing the sectors (electricity, transportation, industrial, commercial, and residential), currently the electricity sector is the largest, representing 40% of total CO_2 emissions. Among all the sectors, comparing the different fossil fuel sources, i.e., coal, petroleum, and natural gas, petroleum constitutes the majority of the emissions at approximately 43%.

It is also suggested that CCRS (Carbon capture-recycle and storage) technologies are highly dependent upon the following four factors:

1. the nature of the application, i.e., a coal-fired power plant, an automobile, air, etc.,
2. the concentration of CO_{2gas} in the gas mixture,
3. the chemical environment of CO_2, i.e., the presence of water vapor, acid species (SO_2, NO_x), particulate matter (PM), etc., and
4. the physical conditions of the CO_2 environment, i.e., the temperature and pressure.

The concentration of CO_{2gas} in the work required for separation decreases as the CO_2 concentration increases. The greater the CO_2 content (i.e. higher chemical potential) in a given gas mixture, the easier it is to carry out the separation (i.e. adsorption).

If the CO_2 concentration in a gas mixture (such as: flue gas; air; gas fuels) is too low, then one has to apply different separation techniques. For instance, in order for **membrane technologies** to be effective, a sufficient driving force (i.e. chemical difference across the membrane) is required. One of the many benefits of membrane technology arises from the act that it is low as regards capital cost. Membranes are a once-through technology in that the gas mixture enters the membrane in one stream and leaves the membrane as two streams, with one of the streams concentrated in CO_2. The chemical-process-based sources of CO_2 tend to have higher concentrations making these processes targets for membrane technology application. Examples include ammonia, hydrogen, and ethanol production facilities.

The chemical environment of CO_2 is important when considering the separation technology since some technologies may have a higher selectivity to other chemical species in the gas mixture. In the process of coal-fired flue gas, water vapor and acid gases (SO_2 and NO_x) will also be present. In a process, which occurs at high temperature or high pressure, it may be possible to take advantage of the work stored at the given conditions for use in the separation process. For instance, a catalytic reaction involving CO_2 may be enhanced at high temperature. It is important to mention that in the case of a catalytic approach for flue gas scrubbing is the case of NO_x reduction to water vapor and N_2 from the catalytic reaction of NOx with ammonia across vanadium-based catalysts. This approach to NOx reduction in a power plant is referred to as selective catalytic reduction. Non-catalytic NO_x reduction, in which ammonia is injected directly into the boiler, is also practiced, but is not as effective as the catalytic approach. Membrane technology is another example, in that molecular separation may be enhanced at high pressure.

The temperature balance of carbon capture has been analyzed. If the process is designed such that the CO_2 separation process (that effectively consumes the thermal energy) runs at the high temperature of the flue gas, then this would be maximizing the thermal content in the system. The flue gas temperature generally is around 650°C at the exit of the boiler, down to approximately 40°–65°C at the stack. Current technologies such as absorption and adsorption are exothermic processes that are enhanced at low temperatures and in a traditional sense are not effective strategies for taking direct advantage of the thermal content of the high-temperature flue gas.

For instance, the capture of CO_2 is most effective at low temperature for absorption and adsorption processes with the regeneration of the solvent or sorbent most effective at elevated temperatures.

- **Membrane separation** and catalytic-based technologies, may however, be enhanced at the elevated temperatures (and pressures) available at exit boiler or gasifier conditions, depending on the specific technology.
- Currently, the largest use of CO_2 is for EOR (enhanced oil recovery). This has been used for a few decades.
- Compression and transport of CO_2. Currently, on average, CO_2-EOR technology provides the equivalent of 5% of the U.S. crude oil production at approximately 280,000 barrels of oil per day. A limitation of reaching higher EOR production is the availability of CO_2. It is reported that some natural CO_2 fields can produce approximately 45 Mt CO_2 per year, with anthropogenic sources slowly increasing (currently 10 Mt CO_2 per year). It has also been suggested that the CO_2-EOR technology (EOR) may also be a useful method to potentially store CO_2.

For instance, the CO_2 used for EOR is not completely recovered with the oil. In fact, only 20%–40% of the CO_2 injected for EOR is produced with the oil, separated, and reinjected for additional production.

Currently, EOR has not had any financial incentive to maximize CO_2 left below the ground. In fact, since the cost of oil recovery is closely coupled to the purchase cost of CO_2, extensive reservoir design efforts have gone into minimizing the CO_2 required for enhanced recovery. If, on the other hand, the objective of CO_2 injection is to increase the amount of CO_2 left underground while recovering maximum oil, then the approach to the design question changes. If there were such an incentive, likely an even larger fraction would stay below ground, via modifications of EOR.

Another study was reported have investigated the co-optimization of CO_2 storage with enhanced oil recovery. Their investigations concluded that traditional EOR techniques such as injecting CO_2 and water in a sequential fashion (i.e., water-alternating-gas process) are not conducive to CO_2 storage. A modified approach includes a well-control process, in which wells producing large volumes of gas are closed and only allowed to open as reservoir pressure increases. In addition to co-optimization of CO_2 storage with EOR, ongoing efforts exist for coupled CO_2 storage with enhanced coal-bed methane recovery (ECBM) and potentially enhanced natural gas recovery.

The post-combustion capture of CO_2 has taken place commercially for decades, primarily for the purification of gas streams other than combustion products.

It is important to recognize that usage of CO_2 in the food industry is not a mitigation option as the CO_2 is subsequently reemitted into the atmosphere, yet the usage of CO_2 continues to drive the advancement of the separation technologies.

It is useful to consider the scale of CO_2 emissions associated with each of the primary energy resources. The annual emissions generated from coal, petroleum, and gas are in the order of 13 Gt, 11 Gt, and 6 Gt CO_2, respectively. Collectively, the emissions associated with the oxidation of fossil-based energy resources are in the order of 30 Gt CO_2 per year. This means that maximum 30 Gt CO_2 is added to the carbon dioxide gas equilibrium in atmosphere, every year. This quantity is reduced, when considering other carbon sinks.

The chemicals produced on a large scale worldwide are as follows:

- Lime, sulfuric acid, ammonia, and ethylene production are in the order of 283, 200, 154, and 113 million tons in 2009, respectively.

4.6   STORAGE OF CO_{2GAS} IN GEOLOGICAL ROCK FORMATIONS

In this geologic cross section, supercritical CO_2 is stored underground in porous rock beneath a layer of impermeable shale. Carbon dioxide does, in some cases, have an economic value, and currently one finds that in the literature research is being carried out on this aspect.

Some such economic aspects are carbonating beverages; usage in materials such as: plastics or concrete; feeding it to plants in enclosed greenhouses; processes of converting CO_2 into methane (CH_4) or liquid fuel.

It is suggested that the large application of CCS, deep saline aquifers, would likely be the major storage site. These reservoirs typically lie some 2–4 km below the surface of the earth, composed of 50-m-thick, porous sandstones filled with saline water. In some parts of earth, these geological formations are estimated to be capable to store few thousands of billion tons of CO_2, sufficient for centuries of emissions. The CO_2 is expected to remain contained in such formations because they lie beneath impermeable shale, and capillary pressure in the sandstone pores holds the CO_2 in place. Over time, the brine reacts with the CO_2 to form solid calcium carbonate (CaCO3).

It is useful to mention that most shells (egg-shells; oyster; etc) all consist of >90% $CaCO_3$ (Chapter 1.10).

4.7   ECONOMICAL ASPECTS OF CCRS TECHNOLOGY APPLICATION

Even though this subject, the economics of the application of CCRS, is out of scope of this book, a few remarks about the economy will be given. In general, pollution monitoring and control and mitigation are expensive for the mankind worldwide. Purification and control of drinking water is one of these examples. Regardless of cost, purification of drinking water being health related has no constricting economic factor. A similar control and mitigation of all kinds are expensive for mankind. Another related, such as corrosion, is one the biggest phenomenon which costs billions for mankind worldwide. The monitoring and control of atmospheric air is thus known for almost a century.

It is estimated that application of CCS to cement industry (BNEF, 2016; Birdi, 2020) could increase the cost of production by 36%–42% for oxy-fuel capture and 68%–105% for post-combustion capture. A reduction in the emissions of the cement could substantially reduce the embodied emissions in a building or house, with a relatively minor impact on the cost of the total construction.

- Fertilizer production: In the production of fertilizers, ammonia is the basis for nearly all synthetic fertilizers globally. With population growth, economic development, and increasing competition for land use there will be increasing demand for productivity increases in food production.

  Presently, most ammonia is derived from hydrogen from fossil fuels and inherent in the hydrogen production process is the stripping out of CO_2. It is suggested that with the application of CCS, the emissions from ammonia production could be reduced by 65%–70%.

- Plastic industries: Crude oil is always treated in refinery. This means that different components of crude oil are separated. After this treatment, suitable technology is applied to produce chemicals for producing a range of plastic products. These different plastic products are extensive:
  - household (buildings, piping, etc)
  - car industry (car parts)
  - infrastructure: roads, sewage, and building

The cracking process in oil refineries is used to produce large amounts of ethylene. Ethylene is a building block for a wide variety of consumer products including plastics, polymers, and detergents. Ethylene is produced from cracking hydrocarbons, usually through steam. CO_2 is produced both through the generation of heat and from the cracking of the hydrocarbons.

Currently, arising from political reasons, various countries (especially Denmark/Norway/etc) are forced to apply carbon capture/recycle techniques. For instance, household waste is sorted into suitable entities, thus managing the recycle process.

Chemicals, steel, and cement industries, all produce CO_{2gas} as a by-product, and CCRS is a major technology to achieve reductions in the associated carbon emissions. Emissions from industry accounted for around 26% of global CO_{2gas} emissions (ca. 10 $GtCO_2$) each year. While there are some emissions in industrial processes which can be reduced through energy efficiency and switching to low-carbon heat and electricity generation, CCS is one of the only options available to address the bulk of emissions generated from chemical reactions inherent in the process of iron, steel, and cement production. It is also the only option available to address emissions from natural gas processing, where CO_2 is stripped from the extracted gas to meet market specifications. CO_2 storage provides an alternative to the current practice of venting this CO_2 into the atmosphere (Bryngelsson and Westermark, 2005).

=======E N D =======. E N. D ======

Appendix A
Climate Change – Basic Conceptions – Carbon Capture Recycling Storage (CCRS)

A.1 CARBON DIOXIDE (CO$_{2GAS}$) SOURCES AND MATERIAL BALANCE (CHEMICAL EQUILIBRIUM)

A.1.1 INTRODUCTION (BASICS)

The purpose of this Appendix A is to address the current man-made activities which may have a (direct or indirect) effect on climate phenomena at equilibrium (such as change in the surface temperature of earth and man-made effect on pollution) (Birdi 2020, Rosenzweig et al., 2021; Lamb, 1995; Gates, 2021).

This climate phenomenon has become very important social (and geo-political) issue for the state of global activities of mankind. It has relationships with the dependent of mankind on: energy sources; food sources; health-medical.

Furthermore, the different parameters involved have become both technical and political aspects (Easterbrook, 2011; Dessler & Parson, 2019; Filho, 2022; Gates, 2021).

In addition, the **chemical evolution,** which has taken place on earth, needs to be considered with respect to the present subject matter (Calvin, 1969). The **chemical evolutionary equilibrium** in (CEE) this step has been recognized as an important parameter (Calvin, 1969; Birdi, 2020; IPCC, 2022; Kemp et al., 2011).

EXAMPLE

SUN
I
I
CO$_{2gas}$ + Water Vapor
I
I
PHOTOSYNTHESIS + GHG
I
I
<PLANTS>

I
<PRE-LIVING SPECIES>
<CURRENT>

<Sun>——-interface(I)
   <I>—atmosphere
      <I>—earth
         <I>—LAND
         <I>—OCEANS

FIGURE A.1   Interfaces at sun heat radiation-atmosphere and environmental conditions. All these are CEE processes.

With the combination of two of these parameters, one finds that the chemical evolutionary equilibrium (**CEE**) has been necessary in this aspect. In addition, at all interfaces, there exists chemical equilibrium (Chattoraj & Birdi, 1984; Birdi, 2020). One finds different CEE examples in environment, such as:

a. photosynthesis (sunshine + carbon dioxide (CO_{2gas})) + water
b. living species (CO_{2gas} -> Plants-> Food-> Metabolism (Living species)-> Lung exhale CO_{2gas})(Food recycle)
c. composition of the atmosphere (different gases)
d. photosynthesis—> plants (pre-living species) + GHG

The magnitude of CEE phenomena, as related to carbon dioxide (CO_{2gas}), is very large, and has been present on earth for billions of years. The creation of living species and environmental thus have been (and are) evolving continuously. Especially, the technology development by mankind has continually been changing the aspects on earth. These developments are expected to effect the future fossil fuel: availability; demand; usage.

Surface chemistry aspects in relation to climate phenomena are found to create *CEE at interfaces*. The Industrial Revolution took place about two centuries ago. The worldwide population has reached about 8 billion during this period. The CEE at interfaces has thus adjusted these different parameters. For instance, the atmosphere – ocean interface is one of the typical phenomena. The phenomenon of CEE throughout the geological chemical evolution, at interfaces in sun-atmosphere-earth system is the current subject. However, it is also recognized that there are dynamically changing sun radiation – atmosphere interfaces. These will depend on the atmospheric conditions: clear skies; cloudy; stormy; tornadoes; etc (Figure A1).

Mankind has induced social control, such as <one child> principal, in China. This led to flattening of population already in 2023. The latter government control has led to various stabilizing effects on:

• world population growth.
• energy demand per capita.

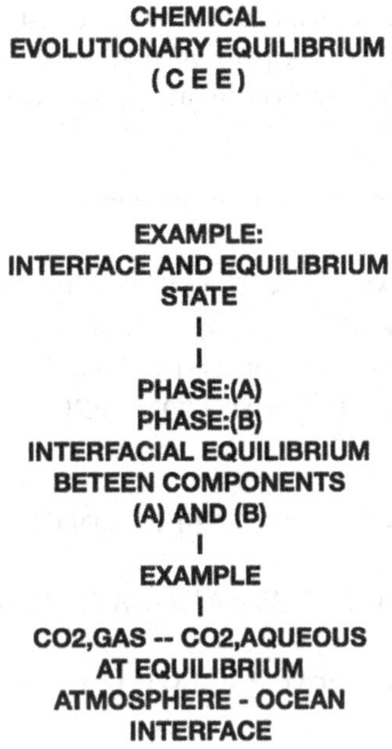

CHEMICAL
EVOLUTIONARY EQUILIBRIUM
(C E E)

EXAMPLE:
INTERFACE AND EQUILIBRIUM
STATE
|
|
PHASE:(A)
PHASE:(B)
INTERFACIAL EQUILIBRIUM
BETEEN COMPONENTS
(A) AND (B)
|
EXAMPLE
|
CO2,GAS -- CO2,AQUEOUS
AT EQUILIBRIUM
ATMOSPHERE - OCEAN
INTERFACE

FIGURE A.2 CEE at interfaces (see text).

- fossil fuel demand per capita: varying throughout the globe: decreasing/stable/increasing.

A few billions years ago, after earth had attained suitable temperatures (i.e. in comparison to current CEE), it is assumed that photosynthesis phenomena were initiated (Calvin, 1969). This means:

- that there was carbon dioxide, CO_{2gas} + sunshine + water.
- that the GHG phenomenon was also present (due to carbon dioxide in air)
- during the CEE, the temperature on the surface of earth has reached an geological constrained equilibrium (i.e. varying from $+50°C$ to $-50°C$).

Any current man-made activity, thus, will affect any of these different chemical equilibria (which have perhaps been active over billions of years). There exists a large gap, between the geological time scale and a short lived man-made innovation. This conceptual difference is being debated in current climate debate (Lomborg, 2022; Gates, 2021.

Living species (insects/animals/mankind) have existed and evolved on earth for many thousands of centuries (Calvin, 1969). Living species are dynamic and varying in many aspects. However, mankind is the only living species which has acclimated to its environment. For instance, mankind is known to have evolved near the equator. But later through adaptations one finds today mankind living almost all over the surfaces of earth. The chemical evolutionary equilibrium (CEE) has developed and living species have survived/existed throughout these thousands of centuries (Calvin, 1969). The latter studies have clearly indicated that *surface chemistry* principles have played an important role in this evolution. The phenomenon of the CEE is found to play a significant role in the evolutionary process of living species, hereunder, mostly for humans.

EXAMPLE:

<CLIMATE-FOSSIL FUEL-ENERGY>
<CLIMATE=EFFECT ON EARTH (SURFACE) TEMPERATURE >
<ENERGY = PRODUCTION USING FOSSIL FUELS >
<FOSSIL FUEL COMBUSTION = PRODUCTION (man-made) OF CARBON DIOXIDE (CO_{2gas})

<*CARBON DIOXIDE* (CO_2) EFFECT ON CLIMATE

<CARBON DIOXIDE + *GREENHOUSE GAS* (GHG)

<GREENHOUSE GAS (GHG) EFFECT ON EARTH SURFCE CLIMATE TEMPERATURE

<RECYCLING AND STORAGE OF CARBON DIOXIDE AFTER CAPTURE (BY: ADSORPTION/ABSORPTION)

<CARBON CAPTURE RECYCLING STORAGE: CCRS

It is significant to mention that the so-called earth surface-atmosphere temperature, as mentioned, is an estimated prediction quantity. In all the current climate models (Lomborg, 2022; Kemp et al., 2022), surface temperature of earth data is used. In other words, one takes an overall average from data. For example, the data from Himalaya and from Sea Deep may be averaged. This procedure has been criticized by others.

For example, the temperature inside the earth is almost constant (lava-like) (about 5,000°C) since the creation about billions years ago (Figure A3).

It also does not include the (nonlinear) gradient of temperature (in the atmosphere or crust). In addition, it is known that the major heat on earth is absorbed (and controlled) by the oceans/lakes/rivers (which comprise 70% of surface area) and oceans with large depths (some places over 5 km deep). The contribution of ocean and atmosphere interface phenomena on climate is complex (Kemp et al., 2022) and very significant.

The magnitude of any temperature measurement at the surface of ocean would be nonlinear with respect to depth (Enns, 2011).

It is recognized that oceans comprise about 70% of earth surface. Further, oceans can reach depths over 5 km. It is therefore accepted that heat absorbed by oceans (and lakes/rivers) is very extensive, in comparison with the rest of earth surface (Filho, 2022; Birdi, 2020; IPCC, 2022) (Figure A4).

```
                            SUN
                          (EMITS
                           HEAT
                       RADIATION)
                             X
                             X
                       ATMOSPHERE
                         (CLEAR)
                        (CLOUDY)
                          (RAIN)
                         (STORM)
                       (TORNEDO)
                             X
                             X
                      E. A. R. T. H
                (LAND.30%)    (OCEANS 70%)
                   (ABSORB/REFLECT HEAT)
                 (OCEANS-FISHERIES-SHELLS-
                         CARBONATES)
                 (OCEANS–OVER 5 KM DEPTH–
                    PSEUDO-EQUILIBRIUM)
```

FIGURE A.3   Sun emits heat radiation (160 million km)-> atmosphere (<10 km) -> earth.
Distance.

CLIMATE MODEL
ESTIMATES

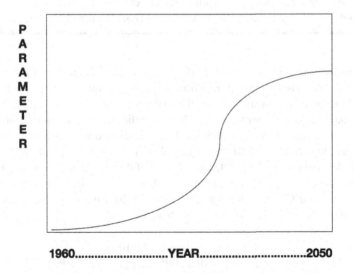

FIGURE A.4   Climate estimate model predictions.

This is in agreement with the current estimates about the fossil fuel emission trends. The current global data are:

- 41 Gton-CO_{2gas} (emissions in 2022)
- 41 Gton peak-CO_{2gas} – (emissions in 2019)

The geo-political development has had a strong effect on the need to usage of increasing the man-made innovation as regards non-fossil energy sources:

- wind/solar energy sources.
- nuclear energy
- other sources: waves/thermal/carbon dioxide recycling

This has resulted in a decisive effect on the decrease in CO_{2gas} emissions; especially in some areas (western Europe). Especially, CCRS technologies are applied in some countries developed.

EXAMPLE:

<GEOLOGICAL-EARTH-ESTIMATED DATA>
    <1>Earth was created 4.6 billion years ago
    <2>Cooling on surface of earth-crust formation: 2.4–2.6 billion years photosynthesis (e.g. Carbon dioxide sunshine water) (pre living plants) (GHG)
    <3>Bacteria from 1.2 billion years
    <4>Higher levels of living species (insects/animals/humanoids)
    <In the current literature, one finds that different climate change *models* are being developed in order to simulate the future predictions (Kemp et al., 2022).

For obvious reasons, mankind is not able to have any effect on all of these geological scale events (i.e. range of billions/millions of years; earthquake eruption; tornadoes). Due to various nonlinear (Enns, 2011) and unknown factors involved, it has been pointed out that these models can only be applicable as estimates (for example: weather forecasts).In these efforts it has been mentioned that such systems, climate change, are known to be immensely complex. The biggest weakness in any man-made model will always be the dearth of scale of data input, as compared to CEE.

Furthermore, all models on climate – carbon (CO_2) – world population – pollution – versus year (about 2100) indicate (Birdi, 2020; Lomborg, 2022; IPCC, 2022; COP27, 2022; Kemp et al., 2022) that at about year 2100:

1. *world population* will stabilize around 12 billion.
2. world usage of *fossils fuels* will stabilize around 2050–2100.
3. world man-made emission of *carbon dioxide* will stabilize or decrease around 2050–2100 <see <2>>.
4. world *surface temperature* will stabilize around 2050–2100 <see 3>.

5. world *energy* production will become less dependent on fossil fuels (and more on non-fossil fuels: wind/solar/hydro/nuclear).
6. GHG phenomena will stabilize (or start to decrease)
7. CCRS and similar technologies will reduce the GHG effect.
8. Non-fossil fuel energy sources will reduce the fossil fuel combustion.
9. the human innovation (using non-fossil fuels) in reducing different carbon dioxide gas addition to the atmosphere.
10. Population of China is already (2023) stagnant (ca. 1.4 billion).

In other words, the task of any meaningful temperature data of earth, on global scale, will be an impossible task. Especially, recent earthquake activities indicate the continuing crust changes (which are mostly unpredictable; erratic). The earthquake effects can be significant considering the dust pollution. Furthermore, lava eruptions give rise to fires and create carbon dioxide, CO_{2gas}.

EXAMPLE:

    <CLIMATE CHANGE (TEMPERATURE RISE) DUE TO MAN-MADE FOSSIL FUEL COMBUSTION
        <TEMPERATURE MEASURMENT OF EARTH
            <ONLY AVERAGE OF SOME FIXED SITES ON THE SURFACE OF EARTH
        <NO TEMPERATURE GRADIENT ANALYSIS
        <TEMPERATURE IN INTERIOR OF EARTH CONSTANT (ca. 6000°C) (DUE TO ITS LAVA-LIKE FLUID STATE)
        <INCONSISTENT SURFACE TEMPERATURES COMPARISON OF EARTH (WITH ATMOSPHERE) AND MOON (WITH NONE ATMOSPHERE)

In addition, it is the intention to indicate, wherever necessary, the effect of natural phenomena which are known to be related to climate change. This Appendix mainly concerns with the system:

- fossil fuels (e.g. coal/crude oil/natural gas) combustion---production of energy + carbon dioxide (CO_{2gas}) production + water.
- Temperature increase in global---due to the GHG property of CO_{2gas}
- Surface chemistry aspects involving carbon dioxide mitigation and control (carbon capture recycling and storage (CCRS)

Currently, much attention has been paid to one parameter, with regard to the effect of GHG (due to different gases in the atmosphere: CO_2; CH4; NOx; etc).

The equilibrium concentration of carbon dioxide in the atmosphere was measured at a given site (Rosenzweig et al., 2021) (Figure 1.11). These data showed an increase in the equilibrium concentration of CO_{2gas} in the atmosphere from 1960 to 2022: from. 200 to 500 ppm. This is in contrast to the fact that:

- any change in temperature measured over land area is not the same as over oceans/lakes/rivers. For example: oceans may be over 5,000 m deep, and conduct heat differently.

It is important to mention that the GHG effect of carbon only relates to when present as gas (i.e. in the atmosphere). The photosynthesis process (plants) existed on earth much earlier than living species or fire (man-made) or industrial revolution was present. This suggests that the sun – atmosphere – earth (land/oceans) system had reached pseudo equilibrium billions years ago.

In other words:

- CARBON DIOXIDE GAS IN ATMOSPHERE =830 Gtons (Gton= 10^9 tons) =GHG
- CARBON DIOXIDE DISSOLVED IN WATER (OCEANS/LAKES/RIVERS)(Carbonate + Shells + Fisheries) – 38,000 Gtons – NON-GHG

These data show that:

- oceans/lakes/rivers are the largest carbon sinks (and have been during the chemical evolutionary equilibrium) (CEE).
- oceans/lakes/rivers are the largest *heat* sinks for the earth.

The same criteria apply to other GHGs which are found in the atmosphere (Chapter 1). However, since none of these gases dissolve in water nor interact with other ions, these exhibit GHG characteristics which are entirely different.

In a particular phenomenon, the fossil fuel combustion has been reported to have a strong impact on climate (Birdi, 2020). The latter is currently being considered as the main phenomena which might affect the climate change, i.e. global (surface) temperature might increase by 1° to 2°C (IPCC, April, 2022).

Accordingly, many man-made processes have been investigated to mitigate this climate change arising from the addition of CO_{2gas} (IPCC, 2022; Lomborg, 2021).

The simple criterion of fossil fuel combustion processes to produce energy, by mankind, is being subject of some concerns and subject to varying conclusions (Birdi, 2020; Lomborg, 2022; Gates, 2022).

EXAMPLE:

<FOSSIL FUEL COMBUSTION
FOSSIL FUEL
$< (C_nC_{n+1}) + OXYGEN (O_2) ==$
$CO_{2gas} + H_2O$

One may discuss this man-made CO_{2gas} from increasing usage of fossil fuel in more detail. This has taken place over the chemical evolution on earth in various stages of time scale. About 200 years ago natural coal/oil/gas reserves were found. Mankind developed technology to use these fossil fuels (always used after being subjected to refineries) for the production of energy; medicine; plastics; etc.

EXAMPLE:

<TIMESCALE (Geological)>
<CURRENT>
<CO_{2gas} IN ATMOSPHERE = 750 Gtons
<CO_{2aq} + Carbonates in Oceans = 48,000 Gtons
<CREATION OF EARTH (ca. 5 billion years ago)>
    <EARTH WAS VERY HOT: THERE WAS NO ATMOSPHERE OR WATER (OCEANS-LAKES-RIVERS)

These data based on geological scale clearly indicate that the initial processes of photosynthesis started only after the appearance of water-age + CO_{2gas} (Calvin, 1969; Birdi, 2020; Kemp et al., 2011).

Based on this aspect, the recycling of carbon dioxide is the main topic in the present context (Birdi, 2020).

EXAMPLE:

>MAN-MADE PHENOMENA<
>CARBON RECYCLING
<1> PURPOSE OF THIS BOOK:
<2>DUE TO INCREASING POPULATION OF MANKIND:
<3>RECYCLING OF CARBON DIOXIDE (PRODUCED BY FOSSIL FUELS COMBUSTION) AFTER SUITABLE CAPTURE PROCESS
<4>CLIMATE AND POLLUTION FROM COLLOIDAL PARTICLES (MANMADE/NATURAL)
<5>MITIGATION OF INCREASING CARBON
>CAPTURE OF CARBON DIOXIDE
    >FROM AIR (420 PPM: 0,042%)
    >FROM ENERGY PLANTS FLUE-SPENT GAS (100,000 ppm: 10% CO_2)
        >RECYCLING OF CAPTURED CARBON DIOXIDE
>ADSORPTION/ABSORPTION OF CARBON
        >NATURAL PHENOMENA EFFECT ON CARBON MASS BALANCE <
        >RAINS/CLOUDS/STORMS/HURRICANES/TSUNAMI/WAR ACTIVITY/EARTHQUAKES/

From this brief description, which shows that in order to decrease the effect of fossil fuel combustion effect on climate, there must be applied useful procedures to control/reduce/recycling of carbon dioxide. It is also recognized that surface chemistry is known to impact this process. This is related to the existence of different surface chemistry principles (Birdi, 2020). Some of the pertinent *interfaces* are indicated in the following.

EXAMPLE:

>DIFFERENT **INTERFACES**
  >SUN HEAT RAYS-ATMOSPHERE INTERFACE
  >ATMOSPHERE-EARTH (LAND) **INTERFACE** 25%
  >ATMOSPHERE-OCEANS/LAKES/RIVERS **INTERFACE** 75%
  <ATMOSPHERE – ANTARICTA **INTERFACE**

It is significant to mention that the ocean surface is very important. Besides, oceans can be over 5 km in depth at some places.

**DIFFERENT INTRFACES IN PRESENT
SYSTEM**

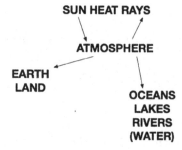

A.2 MASS BALANCE OF HEAT RADIATION RECEIVED FROM SUN

The total magnitude of heat radiation received from sun will be dependent on various paths (Figure 1; Appendix B). The main distinction is that 30% lands and 70% oceans make this aspect significantly important (Figure A5).

A.3 EARTH GLOBAL TEMPERATURE BALANCE

Earth is known to be billions of years old and has formed crust on the surface (due to surface cooling), while the interior is still very hot-lava-like. Earthquakes release some of the hot interior to the surface and cool the planet earth. These phenomena

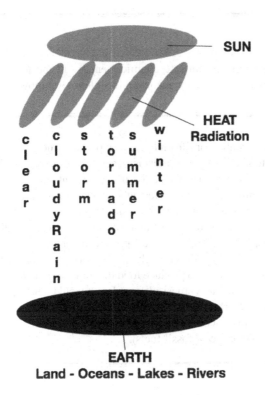

FIGURE A.5 Sun heat-radiation – atmosphere – earth.

are erratic and nonlinear (Enns, 2011; Kemp et al., 2022). The evolutionary cooling of earth, during the crust formation, started some billions of years ago. This initial cooling period is dubious and of obvious reasons not understood completely by mankind.

<NONLINEAR PHENOMENA: all natural phenomena on earth are nonlinear (gravitational movement; temperature variable; wind-rain-hurricanes)

A.4 PHOTOSYNTHESIS EVOLUTIONARY PROCESS

This is the essential evolutionary process, only found since pre-living species, to have been taking place on earth (only in the solar system).

<PHOTOSYNTESIS PSEUDO EQUILIBRIUM EVOLUTIONARY ON EARTH: Photosynthesis has been taking place both during the pre-living species and after-living species. The process is based on pseudo-equilibrium between:

<PHOTOSYNTHESIS>

>**Sun-light + Carbon dioxide** (in atmosphere) + **water** (H_2O).

EXAMPLE:

<**Photosynthesis** is the process by which plants, algae, and some bacteria convert light energy from the sun into chemical energy in the form of sugars (carbohydrates + other essential carbonaceous molecules) (Rabinowitch and Govindjee, 1969).

This process is essential for life on Earth as it provides the energy for plants to grow and produce the food that other organisms depend on for survival. The CEE phenomenon has achieved this which has resulted in the current pseudo-equilibrium of state on the earth environment. During photosynthesis, carbon dioxide from the atmosphere (400 ppm) is taken in and combined with water to produce glucose, a sugar molecule. Oxygen is released as a byproduct of this reaction. This oxygen is then used by other organisms for respiration, which is the opposite of photosynthesis. Photosynthesis is the foundation of the food/fisheries chain, as it provides the energy for plants (on land + oceans + lakes + rivers) to produce food that other organisms (e.g. living species) can consume. Without photosynthesis, life on Earth would not be possible.>

In other words, <carbon dioxide> was essential in the atmosphere throughout the era from pre-living species to present day (after the living species and after the industrial revolution (usage of fossil fuels).

EXAMPLE:

<GEOLOGICAL PHOTOSYNTHESIS STAGES>

<1>CREATION OF EARTH==HOT==NO ATMOSPHERE/WATER/ CO2==NO PHOTOSYNTHESIS

<2> UNKNOWN INTERMEDIATE ERA: Cooling – Crust – Water – CO_{2gas} – Photosynthesis

<3>CURRENT STATE OF PLANET EARTH == Atmosphere (N_2 (70%) + O_2 (28%) + Traces of other gases (CO_{2gas} (420 ppm) + moisture (water)) == PHOTOSYNTHESIS

These observations show that photosynthesis on earth is currently supporting various life supporting parameters.

EXAMPLE

<CURRENT>

<PHOTOSYNTHESIS==CO_{2GAS}(400PPM)+$CO_{2,AQ}$(+CARBONATES IN OCEANS)+WATER

<PRE-LIVING SPECIES>
<PHOTOSYNTHESIS==CO_{2gas}(?PPM)

In all chemical evolutionary systems, it has been found that equilibrium in the system gave rise to sustainable development.

The development in this case maybe:

- growth of plants: trees/food==pre-living
- growth of plants/trees/food==evolution of living species (as today)

These constraints on species necessary for photosynthesis and food and later the evolution of living species, thus, confirm the natural selection of $CO_{2gas} + CO_{2,aq}$.

Thus, the necessary concentration of CO_{2gas} has been subject to control due to the evolutionary process. This requires that the mass of CO_{2gas} in the atmosphere has been maintained by natural evolution through some equilibrium over these billions of years.

<CARBON DIOXIDE EQUILIBRIUM (PSEUDO) IN ATMOSPHERE (System: Sun – Atmosphere – Earth)

The global sun – earth and the current reports about the climate change (on earth) are given much attention in the current literature studies (see ref). These phenomena are known to be dynamic-nonlinear and are given in the following (Kemp et al., 2022).

These sun – earth interaction interfaces have been found to show that these phenomena are both dynamic and nonlinear. In addition, some observations show that these may also be cyclical (Kemp et al., 2022).

EXAMPLE

SUN – (Spots / Orbital Variation)
    I
    I – HEAT – INPUT – VARIABLE
    I
    SUN – **ATMOSPHERE** INTERFACE
    I – *Dynamics*
    I – *Chemical – GHG* – Other
    I
    ATMOSHERE – **EARTH** INTERFACE
    I – *LAND* (25%)
    I
    I – *OCEANS/LAKES/RIVERS (75%)*

ATMOSPHERE – INTERFACE
I
CLEAR SKY – CLOUDY – RAINY
   I
SUN HEAT TRANSMISSION/REFLECTION/GHG
I
PHOTOSYNTHESES ON LAND/ANTARCTICS/OCEANS/LAKES/RIVERS

<DYNAMIC> Sun rotates along some large path in the universe. Earth rotates around the sun (365 days/rotation) and turns around a (varying) tilted axis (24 hours/ revolution). The effect of moon on the earth is considerable.

However, it is useful to briefly describe the planet earth with respect to its climate and the solar system. These form the system which is responsible for any climate characteristics. The relationship of earth climate to surface chemistry is the basis of this manuscript. This latter property is found to be an important aspect of climate and recycling of green + house + gas (GHG), especially CO_{2gas} (Chapter 1).

EXAMPLE:

>SUN – ATMOSPHERE (SURFACE CHEMISTRY) – EARTH (FORESTS-LANDSCAPE-DESERTS-OCEANS-LAKES-RIVERS)

There is a need to mention, briefly, the role and the presence of atmosphere, and the resulting pressure, besides temperature. The overall temperature of earth is dependent on inputs from only two sources:

>HEAT INPUT TO PLANET EARTH <

<source>sun
<source>interior of earth – earthquake eruptions (molten lava-like)

In this (overall) heat input mass balance, sun provides a major continuous(varying) input of heat to the surface of earth . On the other hand, the input of heat from the interior of earth to its surface is also found to be erratic. For instance, when hot lava during an *earthquake* erupts to the cooler crust of earth. This has a cooling effect on the global temperature of earth. The heat dissipates into the space. However, the current reference to climate change: rise in temperature: this elates to only change in temperature on the surface of earth (at some fixed locations).

Any atmospheric pressure on the surface of earth is related to the molecular pressure arising from the gas molecules in the atmosphere. This is defined as one unit of bar, at the sea level.

This may indicate that gases which were initially released from the earth, during its early creation, were attracted to the earth and thus were responsible for the formation of atmosphere (equilibrium state current). These gas molecules thus formed the original atmosphere. In gas phase, contrary to other phases (e.g. liquid; solid), all molecules mix homogenously. The latter may or may not be the case in other phases.

In other words, these gases could not escape the gravitational attraction from the earth. It is also found that the atmospheric pressure drops on earth as one moves higher (Rosenzweig et al., 2021).

In addition, the complex system:

$$PHOTOSYNTHESIS \ (ON \ LAND/IN \ OCEANS\text{-}LAKES\text{-}RIVERS)$$
$$= SUN + CO_2 + WATER$$

The process of photosynthesis is unique as regards the sun – earth in the solar system. Earth is known to be the only planet where photosynthesis processes can occur due to appropriate conditions, as determined by the chemical evolutionary processes (Calvin, 1969). The process of photosynthesis on earth is known to have been active long before living species appeared. In other words, CO_{2gas} has existed during this photosynthesis period (pre-living species).

It is useful to briefly consider the phenomena of *chemical evolution and climate change.*

The chemical evolution studies have shown details on this phenomena (Calvin, 1969; Birdi, 2020). The latter factors are:

- sun shine
- carbon dioxide and water

In addition, the chemical evolution has profound effect on all kinds of living species, dependent on food.

Every planet in the solar system is known to exhibit different climatic characteristics: such as:

- atmosphere: non-present
- atmosphere; present
- composition of gases in atmosphere;
- structure of planet (surface and interior properties);
- temperature of planet (surface and interior).
- distance from the sun
- photosynthesis (only present on earth)

Most importantly, living species (especially humanoids) only are known to exist and to have survived on earth over many hundreds of thousands (over 250,000) of years (Birdi, 2020; Calvin, 1969; Kemp et al., 2022). Furthermore, humanoids are the only living species who have been able (only to some extent) to change or effect the natural surroundings (including climate variations). For example, humanoids have had no effect/control on:

- earthquakes
- storms/hurricanes/flooding
- ice age

EXAMPLE:

>LIVING SPECIES (ON EARTH) AND FOOD
>FOOD – PHOTOSYNTHESIS

It is also evident that all living species are dependent on food (photosynthesis = plants/ fisheries), which the chemical evolution has created on the earth (Calvin, 1969; Birdi, 2020) throughout its creation.

The chemical evolution and the various evolutionary equilibriums have been the determining factors. Especially, the chemical equilibrium has maintained the parameters under discussion:

- cooling of earth surface to form crust.
- photosynthesis phenomena based on the availability of sunshine + CO_{2gas} + water.
- support of photosynthesis = plants (food/fisheries) for the evolvement of living species.

It is important to mention the latter, as these are known to effect the temperature of the earth (climate change).

The main chemical criteria of the latter observation have been related to the importance of carbon dioxide (CO_2) (Birdi, 2020). Besides various other properties, the current GHG property related to carbon dioxide and its increasing production from fossil fuel combustion has been correlated with climate change (i.e. increase in the temperature of planet earth). However, there exist other gases in the atmosphere with GHG properties, e.g. methane and water vapor. The role of stable and equilibrium (pseudo) substances in the chemical evolution has been found to be important.

EXAMPLE:

>CHEMICAL EVOLUTION AND CARBON DIOXIDE AND PHOTOSYNTHESIS AND FOOD AND LIVING SPECIES

This has mainly been correlated with man-made increasing input of carbon dioxide into atmosphere (Birdi, 2019; Kemp et al. 2022).

Furthermore, currently there has been concerted efforts by various international agencies on climate change (such as: United Nations Framework Convention on Climate Change: UNFFCCC).

There have been activities over the past few decades with respect to increasing production by man-activities. Despite these efforts, the concentration of carbon dioxide has been observed to be slowly increasing (in air). Based on these efforts, some of the climate models indicate that around 2100, the rise in temperature (due to GHG effect from carbon dioxide) will be between 2.1°C and 3.9°C (Kemp et al., 2011).

EXAMPLE:

>GLOBAL EARTH TEMPERATURE INCREASE BY YEAR 2100 MUST REMAIN UNDER 2.1°C–3.9°C (surface) INCREASE<

There are also literature data which do not completely agree with this climate model (deny) forecast. This aspect of climate subject is out of scope of this context.

Especially, the mass balance of carbon dioxide (in gaseous state) requires analysis of all states where this substance is found. In the following, this analysis is delineated. The temperature change data are estimates, since any natural phenomena (such as: earthquake/rains/hurricanes/pollution) have not been considered. The latter phenomena are erratic and thus would be impossible to predict and inclusion in any model. Hence, it is useful at this stage to consider the mass-balance and equilibrium of carbon dioxide gas, as found in the free state in different phases. In the present case, only carbon dioxide is considered, since it is considered to be most prominent (Birdi, 2020).

EXAMPLE:

<MASS BALANCE OF CARBON DIOXIDE   (G A S) AROUND EARTH >

(CARBON DIOXIDE (GAS) AS PRESENT IN AIR IS KNOWN TO BE BOTH IN EQUILIBRIUM WITH OTHER PHASES)

  I. $CO_{2,AIR} = CO_{2,AQ}$ (OCEANS/LAKES/RIVERS) (FAST).
       EQUILIBRIUM CONSTANT:
       $K_{CO2} = [CO_{2air}]/[CO_{2aq}] = $ ca.750 Gtons/35000 Gtons
 II. $CO_{2,AIR}$ IS ADSORBED BY PLANTS GROWTH (SLOW) (PHOTOSYNTHESIS)(SUMMER ONLY)
III. $CO_{2,AIR}$ IS PRODUCED BY FOSSIL FUEL COMBUSTION
 IV. $CO_{2,OCEANS}$ IS USED BY FISHERIES

ATMOSPHERE
CO2,gas

OCEANS/LAKES/
RIVERS
CO2,aq + H2O = H2CO3
H2CO3 = HCO3- + H+
HCO3- = CO3-- + H+
CATIONS (CALCIUM/MAGNEIUM) +
CO3-- = CARBONATES
Ca++ + CO3-- = CaCO3

(Systems where either carbon dioxide or related pollution substances is involved and those which are natural and erratic (out of control of mankind))

A. EARTHQUAKE ERUPTIONS (WHICH GIVE RISE TO:BOTH CO₂ AND POLLUTION)
B. WARS (LARGE FOSSIL FUEL COMBUSTION AND POLLUTION)
C. POLLUTION RELATED TO DIFFERENT SOURCES (EARTHQUAKE; WARS; GLOBAL INFECTION)

These processes show that it is insufficient and inaccurate to consider only the state of concentration of carbon dioxide in air (Birdi, 2020).

EXAMPLE:

<CARBON DIOXIDE (CO_{2gas}) SINKS
(SINCE CO_{2gas} IS SOLUBLE IN WATER)

Especially, on a long geological term basis, these equilibria will change and thus affect any long term models (emic).

However, it has been reported that almost three decades have been used by the United Nations Framework Convention on Climate Change (UNFCCC) to control/ mitigate the magnitude of GHG emissions. Regardless, the concentration of anthropogenic GHG emissions tends to slowly increase. These models are procedures to achieve temperature change of about two degrees centigrade. However, similar climate related models predict that world population will stabilize around 2059 (at 11 billions).

The phenomenon of photosynthesis is the most essential life sustaining system Thereafter, all substances which are needed for photosynthesis (e.g. CO_{2gas}; water; oxygen) are thus life sustaining substances. (.).

EXAMPLE:

>CARBON DIOXIDE (CO_2) (IN AIR: 400 PPM) AND LIVING SPECIES ON EARTH
>All plants/food/fisheries are mainly made from photosynthesis (e.g. sun-carbon dioxide-waters)
>All carbonaceous substances are made from carbon dioxide (air)
>All carbonaceous substances in oceans are made from carbon dioxide (as dissolved: $CO_{2,aq} + H_2CO_3 + CARBONATE\ SALTS + FISHERIES$)

EXAMPLE:

<GLOBAL CO_{2gas} MASS BALANCE EVOLUTIONARY EVIDENCE
…..EVOLUTION STAGE……………………ATMOSPHERE

<I>CREATION (PRE-WATER-CO_2)…………………….NO

<II>CRUST (PHOTOSYN (WATER-CO_2))………..YES (CO_2 pre fossil fuel)

<III>CRUST (PHOTOSYN+FOSSIL FUEL)…..YES (CO_2, fossil fuel)

The concentration of carbon dioxide, in the atmosphere, is thus found to be in equilibrium throughout the evolution of:

<photosynthesis – plants – food + fisheries--

Additionally, the extensive activities related to the living species only on earth give rise to increasing degree of pollution of different kinds. For example, alone population growth of humanoids is doubled from 4 billion to 8 billion over the past 50 years. It is also obvious that the contribution to pollution and climate will thus be related to the latter.

This phenomenon is important in the present context. This arises from the fact that the increase in technical activity on earth is directly related to the number of mankind. Although, due to some social constraints, the world population is predicted to flatten out in a few decades (ca. 11 billion) (Chapter 1). This means that the state of various man-made activities would reach a state of equilibrium.

EXAMPLE:

<STABLE WORLD POPULATION-ENERGY NEEDS-CLIMATE INTERACTION

This is reported to be related to the following:

- increasing fossil fuel usage (combustion) with increasing worldwide population and industrialization.
- increasing production of different types of pollutions (e.g. air-pollution; drinking water pollution)
- climate change (on earth) related to natural phenomena (earthquakes wars; etc)
- climate change related to different kinds of pollution: air pollution; water pollution; dust pollution; cloud formation; internal heat balance (inside of earth is very hot: lava-like: 7,000°C)

In the case of the planet earth, these phenomena can be described as follows:

- atmosphere. sun interacts with gases /GHG)/clouds in the atmosphere before its rays reach the earth (land/oceans (75%))
- earth atmosphere and sun: earth is the only solar planet with atmosphere (air) which is found to regulate the planets temperature (compare this observation to another planet without any atmosphere)
- crust (i.e. surface) of earth with green fields-deserts-mountains-oceans (75%)-lakes-rivers-snow covered (Antarctic-north pole)
- interior of earth: hot-lava-like.

SUN (HEAT EMISSION)
>ATMOSPHERE (CLEAR DAY)
>ATMOSPHERE (CLOUDS)
>ATMOSPHERE (POLLUTION A
>ATMOSPHERE (LAND OR OCEANS/LAKES/RIVERS)

These phenomena, somewhat complex (and with various unknown parameters), interactions show that not only these phenomena are nonlinear, but also vary from place to place. Especially, any unpredictable natural phenomena, such as: earthquake; war; epidemic, would lead to incorrect man-made-model conclusions.

Climate (especially temperature) has been found to be changing constantly over many long and short intervals (geological time scale). The phenomenon of earthquake convincingly indicates that the planet earth is dynamic and its overall temperature, as regards heat input from its interior, is cooling. In other words, the hot lava from interior is cooled when reaching the crust. The tilt of earth axis has been reported to have had a significant effect.

The intermittent and continuous eruption of hot lava from the molten interior is a direct evidence of the dynamics. It is also obvious that any such analysis would be subject to very large approximations. In other words, any earthquake activity with large scale eruption will thus be non-predictable and will not be accounted for by any useful climate modeling (Kemp et al., 2011).

This arises from the fact that the system under consideration is very large and varying (Appendix A). For example, mankind has for obvious reasons no control over the earthquake activity (related to the inner parts of the earth). A similar mention can be made for other natural phenomena: storms/floods/hurricanes/tides.

The geological aspects will be out of scope in this study. However, a very concise mention will be given for the sake of clarity and simplicity.

Most importantly, a short description of geological aspects of climate on earth needs to be indicated:

- climate on the surface of earth is what mankind experiencing and this is related to common reference (as used in the current context)
- climate change is common related to change

<SOLAR HEAT FLUXES (ESTIMATES)(KIEHL AND TRENBERTH)
     SOLAR EMISSION>>>342 W/m^2
     SOLAR RADIATION REFLECTED<<<107 W/m^2
     SOLAR REFLECTED BY ATMOSPHERE<<<77 watt/m^2
     SOLAR TOTAL EMISSIONS FROM EARTH<<<390 watt/m^2
     SOLAR TOTAL ABSORBED<<<350 watt/m^2
     SOLAR REFLECTED BY SURFACE<<<30 watt/m^2
     SOLAR ABSORBED BY EARTH SURFACE<<<168 watt/m^2
     SOLAR RADIATION ABSORBED BY ATMOSPHERE<<<67 watt/m^2

A.5   CARBON DIOXIDE (CO_2) CYCLE IN OCEANS/
LAKES/RIVERS (ESTIMATED)

In the lack of exact data, one may resort to estimates based on comparable references. This approach has been made in conjunction with the carbon dioxide solubility in water (as found on earth: oceans/lakes/rivers). These comprise different states of CO_2 in this rather extensive system.

EXAMPLE:

     <CARBON DIOXIDE (CO_2) CYCLE INVOLVES ONLY
     <

The main reason is then briefly considering the age of the solar system and the planet earth (estimates).

EXAMPLE:

CLIMATE (GLOBAL) VARIATION ON EARTH
>BEGINNING: EARTH AS HOT LAVALIKE (very hot-pre water (oceans –
lakes – rivers – plants + living species)

>TODAY: EARTH CONSISTS OF DIFFERENT PHASES:
INTEROR: HOT AND LAVA-LIKE
CRUST UPPER STATE:
LAND AREA
OCEANS (75%)
ATMOSPHERE

>THIS CONVINCING SHOWS THE CHANGING EARTH TEMPERATURE
<

All these various sun – earth heat radiation interactions are known to be nonlinear estimations (Enns, 2011).

This thus makes any simple modeling or short-/long-scale predictions questionable and inadequate (Lomborg, 2022). The main criticism has been that modeling requires long age data (which are obviously not available) for input. In addition, the nonlinearity (as present in all natural phenomena) makes these inaccurate. In addition, the erratic input arising from natural phenomena, such as: war activities/earthquakes/storms/hurricanes, is not included in these models.

Climate temperature-change-GHG models:

Currently, various efforts are devoted to correlated-predictions-control about the effect of any man-made effect on climate (especially effect on global earth temperature). All these considerations are obviously out of range of this study. However, there are some areas which do interact with some criteria with this subject content.

These different models are varying and for obvious reasons elaborate. However, some simple phenomena as already mentioned herein are useful items.

EXAMPLE:

    <CLIMATE MODELS WITH LIMITATIONS >
    <1> CLIMATE TEMPERATURE MEASUREMENT
  <TEMPERATURE IS MEASURED ON THE SURFACE OF EARTH – AS
REFERENCE
    <EARTH TEMPERATURE VARIES
xxx

EXAMPLE:

Following is estimated about the masses of fossil fuels and energy in the USA:

<it is estimated that in the year 2006:

<U.S. consumed about 22 trillion cubic feet ($22\,m^3$) of natural gas (21% of the world total), compared to 105 trillion cubic feet ($18\,m^3$) worldwide.

<U.S. produced about 18 trillion cubic feet compared to about 105 trillion cubic feet worldwide.

<US is reported to have about 22% of the energy used in the homes from natural gas.

(from about 425,000 natural gas well sites.

However, living species are known only to have existed on the planet earth (Calvin, 1969; Birdi, 2020) in the solar system. Among these species, mankind is the only variety which had evolved in different ways and is continuously evolving as evidenced by different criteria. Mankind has evolved and has had some effect on the environment and thus also on climate. These mankind activities thus may be used in future to make any mistakes of the past.

Accordingly, mankind, in recent decades, has both analyzed and taken action in controlling and mitigating any effect that climate may have on the existence of living species. For example, the use of Freon was limited in order to control the ozone-atmosphere layer (Birdi, 2020). Recently, the possible effect of increasing use of fossil fuels by mankind has been reported to have been:

- pollution (all kinds)
- production of carbon dioxide (CO_2) has been of active concern around the globe (Birdi, 2020).

Accordingly, humankind has mitigated fossil fuel consumption by increasing the use/replace by non-fossil fuels:

(DIFFERENT NON-FOSSIL)

- geothermal
- hydro
- nuclear
- wind
- solar
- wave
- carbon dioxide capture (adsorption/absorption) and storage
- etc: (electro-chemical)

In the present context, the subject matter is the consideration of climate change and temperature and its relation to man-made activity; especially after the industrial revolution (Birdi, 2020). This is specifically related to increasing fossil fuel technology (combustion) and thereby increasing amounts of pollution (Birdi, 2020), and different gases (as by-product) release into atmosphere. In addition, it is also observed that the mankind population has increased approximately 4-fold every 50 years. However, it is projected that this phenomena will level off around 2050–2100 (Figure 1.7).

EXAMPLE:

MAN-MADE FOSSIL FUEL COMBUSTION
(Current usage: oil – 100 million barrels/day; natural gas – 30 million equivalent; coal – 30 million equivalent + other energy sources (hydroelectric energy, geothermal energy, atomic energy, wind energy, and solar energy)
Fossil fuel combustion > energy + pollutants (gases: carbon dioxide, NOx, SO_3, and soot-particles)

It is thus also important to notice that air-pollution and any climate change (related to temperature/storms/hurricanes/etc) are inter-related and will be expected to be nonlinear (Enns, 2011).

- Pollution and fossil fuel related industrial activity: A brief mention on the pollution phenomena is useful in the present context. Mankind has, over the past many decades, noticed that pollution of different kinds may affect the life conditions. For example, it has been noticed that drinking water must be treated properly, such that it is free of bacterial infections. However, this has not been treated equally at different parts of world. In the same manner, large cities are known to produce local pollution resulting from different technical developments: such as:
  - traffic
  - large events (arenas; meeting halls; festivals)
  - treatment of sewage
  - clean air processes

In this aspect, it is noticed that humans have not treated pollution in different conditions. For instance, only a few decades ago un-treated sewage waste was dumped into oceans worldwide. Exhaust gases from vehicles or factories are entering the atmosphere without complete and necessary treatment.

Man-made pollution from traffic (cars/trucks/air) has effect on climate change.

EXAMPLE:

    CARS
    SHIPS
    PLANES
    CEMENT INDUSTRY

However, even though the concentration of carbon dioxide (CO_{2gas}) is comparatively low (0.04%–400 ppm), it plays a very significant role as regards the existence of living species/plants/food/fisheries on planet earth ( ). In fact, the magnitude of the concentration of carbon dioxide has been found to have a significant role.

Additionally, the concentration of carbon dioxide in surroundings of earth needs to be considered here. Earth is surrounded by the atmosphere and as well as its 75% is covered by oceans/lakes/rivers. The carbon cycle in its various states has been estimated (with large inaccuracies as expected from such systems related to

---

EXAMPLE:

CO_{2gas} == 750 GtC
   CO_{2aq} == 39,000. GtC
   <CHEMICAL EQUILIBRIUM (KCO_2,aq)
   KCO_2,aq = 750/39,000
          = 1:50
   (where: CO_{2aq} -> oceans/lakes/rivers)

These equilibrium ($K_{CO2,aq}$) data show that approximately when 50 GTC is added to air, then only 1/50 GTC is added to air, while 49 GTC is added to oceans/lakes/rivers. These quantities are estimates and will be expected to be different from real values. This arises from the fact that activities are needed.

This human technology development needs to be mitigated with respect to both positive and negative effects on living species and climate environment.

However, the pertinent geological and natural development one needs to mention is that living species are only found on planet earth (based on the composition of substances and as well as the range in variation in temperature and pressure: as found in different parts of earth). At the same time, it must be noted that the extent of pollution from fossil fuel combustion is separate phenomena.

Planet earth in the solar system has various physico-chemical properties which have determined the state of non-living and living species on this planet. After planet earth was created and was a very hot ball (billion years ago), the temperature started to cool (for some unknown reasons) in the surroundings: A solid crust was formed.

EXAMPLE:

MAIN GEOLOGICAL STATES OF EARTH:
   >ORIGIN....…..BILLIONS OF YEAR AGO....VERY HOT
   >COOLING PERIOD – CRUST FORMATION – INTERIOR VERY HOT –
PRE-LIFE – PRE-WATER
   >LIVING SPECIES ERA – WATER-PLANTS – FISHERIES – FOOD--
FORESTS –PHOTOSYNTHESIS ($CO_2 + H_2O + O_2$)
   >PRESENT ERA – LIFE (MANKIND/LIVING SPECIES)
   >

In other words, there was a geological period pre-life (pre-living species) and later (until present) with pro-life era. The states of these evolutionary stages have been described in the literature (Calvin, 1969; Birdi, 2019).

This analysis shows that earth has been under constant change since the creation of the earth: and thus the climate may also be expected to be continuously changing.

During the cooling-period, this leads to the irregular crust formation on the surface, while the interior still remains fluid and hot (very high temperature). The lava eruptions under earthquake show this phenomena. This evolutionary cooling at the surface of earth shows that temperature has been (and still is) changing. However, this phenomenon is considered as nonlinear-complex-unpredictable ( ). Besides, this is dependent on time/place on earth. This geological event is much earlier than the so-called lesser warming era: ice-age (about 15,000 years ago), only observed in the northern hemisphere (. 9). These considerations also clearly indicate that on a geological time scale, some degree of cooling is taking place.

EXAMPLE:

EVOLUTIONARY TEMPERATURE OF EARTH:
>FROM BEGINNING::::FLUID AND VERY HOT ::PRE-LIFE ERA
>SURFACE-CRUST FORMATION STATE::::COOLING ON SURFACE WITH LAND/MOUNTAINS:::HOT ::PRE-WATER
>FORMATION (APPEARANCE) OF WATER (CLOUD FORMATION) AND OCEANS::COOLING PERIOD

In the above example, only concentration of carbon dioxide (CO_{2gas}) is given. This is due to fact that the current subject deals mainly with the effect of change in concentration of CO_{2gas} in the atmosphere on earth. Furthermore, life on earth is mainly dependent on the continuing need of carbon dioxide (CO_2) for the production of food/plants (photosynthesis being the main/essential requirement for living species). Most biologically metabolism of living species are based on oxidation of food-intake and production of energy + production of useful carbon based growth processes + carbon dioxide (exhaled as CO_{2gas}).

Additionally, the evolutionary phenomena also support the fact there is need for control presence of CO_2 (as gas at rtp) in air (Birdi, 2020). There is an increase in fossil fuel combustion and concurrently:

- increase in production of various polluting gases: $CO_2/CO/SO_2/NOx$.
- increasing degree of pollution (particles in air; etc).

The sun – atmosphere (air: nitrogen; oxygen; water-vapor; GHGs (methane; NOx; carbon dioxide)) interactions are currently being analyzed and evaluated under different circumstances (Birdi, 2020).

This geological chemical evolution is not very clear, for obvious reasons. As regards the living –species, continuous (over geological era) existence is convincing evidence for natural equilibrium. The latter equilibrium in natural systems thus has quasi-stabilized the systems on planet earth.

However, it is useful to compare two different sun-planets and the surface temperature:

- sun and earth (with atmosphere)
- sun and moon (almost no atmosphere)

It is found that the surface temperatures are drastically different in the two cases.

EXAMPLE:

SURFACE TEMPERATURE OF EARTH AND MOON
  >EFFECT OF ATMOSPHERE (ONLY)

PLANET	DAY	NIGHT	ATMOSPHERE
MOON	127°C	−173°C	none
EARTH	ca. +50°C	−70°C	ATMOSPHERE

The fluxes of carbon dioxide (CO_2) production and influxes in the atmosphere have been estimated in recent studies (Birdi, 2020).

However, for obvious reasons these are estimates and are not very reliable. Especially, most of the phenomena are known to be nonlinear.

However, the temperature balance was eventually established where the hot planet earth was undergoing various chemical changes. These were mostly related to the living species which could survive under the given conditions. Some literature studies have been reported on these evolutionary aspects (Calvin, 1969; Birdi, 2020).

EXAMPLE:

Evolutionary diagram of planet (earth) schematic):
    Begin (formation of earth/ sun):
    <1>Sun.....Earth (hot-fluid)(fluid-lava like)
    <2>Pre-life: Earth: Hot with crust (pre-water)(pre-photosynthesis)(high temperature-pre-living species)
    Life on the surface of earth:
    Appearance of cells and later larger living species: current living species.

The subject matter related to climate (temperature: in the present context) change related to earth (and mankind) is rather complex (Calvin, 1969; Kemp et al., 2022; Birdi, 2020). This has also been the reason that no simple criterion has been suggested or pursued. This is becoming more and more obvious from the results as one finds at the international meetings (such as: COP26).

EXAMPLE:

CLIMATE SURROUNDING THE EARTH
   Climate term is used to express the state regarding:
   ...Wind/Waves on oceans/
   ...River/ocean floods
   ...Temperature (an ambiguous quantity)
   ...Tornadoes /Storms/Monsoon

On the other hand, the impact of concentration of carbon dioxide (CO_2) is not only on climate (temperature) change, but also on other daily life essential factors.

In fact, CO_{2gas} (carbon dioxide) generally is termed as gas of life (especially food: e.g. on land (wheat/corn/vegetables/fruits) and oceans (fisheries: e.g. fish (skeletons + shells + etc).

In fact, this indicates that evolutionary era of living species has shown this stage:

- sun – oceans (fisheries)

One of the most significant substances in relation to life on earth is carbon dioxide (CO_2) in the gas form. The molecule CO_2 is essential for many life essential and sustained process. The photosynthetic role of CO_2 in the growth of plants and food is obviously the most significant beginning step. It has been reported that this took place before life appeared (evolved) on earth (Calvin, 1969). All carbonaceous substances as found in living species on earth are dependent on carbon dioxide in the atmosphere. This is in spite of the fact that the concentration of the latter is around 400 ppm (0.04%).

Furthermore, only recently (about 15,000 years ago) the ice-age was warming in the northern hemisphere (..). This shows that earth was undergoing a warmer geological period (and it may still be undergoing this process).

In all cases, the climate and temperature are dynamic systems and are not attributed to any simple parameter.

EXAMPLE:

@ PHENOMENA OF PHOTOSYNTHESIS
SUN-RAYS REACHING THE EARTH – PLANTS + CO_{2gas} + WATER (H_2O)

As mentioned elsewhere, the concentration of CO_{2gas} becomes critical as it is a lethal gas at concentration above approximately 1% (10.000 ppm) (Appendix B).

The evolutionary growth of life living species on earth thus shows that concentration of CO_2 in air has been maintained under this threshold for life to sustain throughout the evolution.

The criterion called greenhouse gas (effect) (GHG) has been noticed with respect to the increase in both CO_2-air and temperature surrounding the earth (Birdi, 2020; Lomborg, 2022; Kemp et al., 2022).

The subject matter discussed here is mainly related to educational aspects of climate + temperature change on or around earth.

The purpose of this Appendix is to add some more basic educational information on the carbon capture recycling and storage CCRS technology as described in Chapter 4. CCS technology is expanding rapidly and there are many different aspects which need to be addressed. Some additional data are mentioned here for those readers who may be interested in pursuing further studies, based on the online approach. The current state of the criteria on climate-change and any man-made effect needs to be evaluated.

>GLOBAL AVERAGE SURFACE TEMPERATURE DATA OF EARTH
(average of earth surface)

In the case of consideration involving any change in surroundings of earth and different man-made factors needs a basic understanding. These factors are used in this Appendix.

Current subject is related to any climate/temperature-change. The average temperature of earth is not easy to define or measure. A more simplistic approach has been used. The general procedure is that some chosen sites around the world on the surface of earth are used for monitoring the average temperature. A paradigm has been used to estimate an average temperature of earth from these data (IPCC, 2005, 2007, 2011; Lomborg, 2007, 2022).

EXAMPLE: Size of atmosphere surroundings the earth.

The earth is surrounded by a layer of atmosphere (consisting of air: composed of different gases: nitrogen (78%)/oxygen (21%)/CO_2 (0.04 %)/traces of other gases). Gas molecules in the atmosphere are attracted to the earth through van der Waals forces. This force decreases with the height from the surface of earth. It is found that at about 20 km from the surface of earth there is almost no atmosphere and mankind cannot survive under normal conditions.

In the following some typical temperature data are given (Mathews et al., 2009). These are very few and only point to the complexity of these data. The data include for land and sea over the year 2014 (Birdi, 2020). Some important features may be mentioned.

Ranking order is (year/degree):

1.	2014	0.6
2.	2010	0.6
3.	2005	0.5
4.	2007	0.5
5.	2006	0.5
6.	2013	0.5
7.	2009	0.5
8.	2002	0.5
9.	1998	0.5
10.	2003	0.5

>>>A.
SUN AND THE TEMPERATURE ON EARTH (only surface)

The sun (star in solar system) is on average 100 millions of miles (160 million km) away from the earth. It is almost a million times bigger than earth (by volume). The solar energy reaching the earth is varying with time and place. It is found that there exists a 11 year solar activity cycle. Besides, while the earth rotates (24 hours/rotation) and travels an elliptical path (over 365 days/revolutions) around the sun, the heat input from sun is varying constantly. This gives rise to climate variation which is very complex to any observers.

EXAMPLE:

SUN........100 MILLIONS   MILE......(ATMOSPHERE).......EARTH

All kinds of interactions between the sun and planet Earth are known to be nonlinear in physical terms. The nonlinear and erratic interaction are also due to the atmospheric characteristics (Birdi, 2020).

This phenomenon is also found to indicate that any simple modeling can be fruitless and inaccurate (Birdi, 2020).

The Sun provides the primary source of energy (heat) driving Earth's climate system, but its variations have played very little role in the climate changes observed in recent decades. Direct satellite measurements since the late 1970s show no net increase in the Sun's output, while at the same time, global surface temperature measurements have shown increases. For earlier periods, solar changes are less certain because they are inferred from indirect sources – including the number of sunspots and the abundance of certain forms (isotopes) of carbon or beryllium atoms, whose production rates in Earth's atmosphere are influenced by variations in the Sun spots.

>Solar activity cycle:

Data show that the 11 year solar cycle, during which the Sun's energy output varies by roughly 0.1%, can influence ozone concentrations, temperatures, and winds in the stratosphere (the layer in the atmosphere above the troposphere, typically from 12 to 50 km, depending on latitude and season).

These stratospheric changes may have a small effect on surface climate over the 11 year cycle. However, the available evidence, however, does not indicate pronounced long-term changes in the Sun's output over the past century, during which time human-induced increases in CO_2 concentrations have been the dominant influence on the long-term global surface temperature increase. Further evidence that current warming is not a result of solar changes can be found in the temperature trends at different altitudes in the atmosphere).

The measurements of the Sun's energy incident on Earth do not show any increase in solar flares during the past 30 years. The data show only small periodic amplitude variations associated with the Sun's 11-year cycle. Some results from mathematical/physical models of the climate system showed that human-induced increases in CO_{2gas} would be expected to lead to gradual warming of the lower atmosphere (the troposphere) and cooling of higher levels of the atmosphere (the stratosphere). In contrast, increases in the Sun's output would warm both the troposphere and the full vertical extent of the stratosphere. At that time, there were insufficient analytical data to test this prediction, but temperature measurements from weather balloons and satellites have since concerned these early forecasts. It is now known that the observed pattern of tropospheric warming and stratospheric cooling over the past 30–40 years is, in general, consistent with computer model simulations that include increases in CO_2 and decreases in stratospheric ozone, each caused by human activities. These data are not consistent with purely natural changes in the Sun's energy output, volcanic activity, or natural climate variations such as El Niño and La Niña.

Despite this agreement between the global-scale patterns of modeled and observed atmospheric temperature change, there are still some differences. The most noticeable differences are in the tropical troposphere, where models currently show more warming than that has been observed, and in the Arctic, where the observed warming of the troposphere is greater than that in most models. The observed data on the warming in the lower atmosphere and cooling in the upper atmosphere provide some useful information on the reasons of climate change.

>>

§Climate Change at the surface of earth (GENERAL REMARKS):

In this Appendix, a short description is given to describe some studies as related to climate (as regards (global) temperature) of earth. The past observations indicate that the climate on earth is variable over short and long time scales. The most obvious is the ice age which led to the melting of ice around Northern Europe (some 15,000 years ago).

The largest global-scale climate variations in the Earth's recent geological past are the ice age cycles, which are cold glacial periods followed by shorter warm periods. The last few of these natural cycles have recurred roughly every 100,000 years. They are mainly paced by slow changes in Earth's orbit which alter the way the Sun's energy is distributed with latitude and by season on Earth. These changes alone are not sufficient to cause the observed magnitude of change in temperature, nor to act on the whole Earth. Instead, they lead to changes in the extent of ice sheets and in the abundance of CO_2 and other greenhouse gases which amplify the initial temperature change and complete the global transition from warm to cold or vice versa.

Recent estimates of the increase in global average temperature since the end of the last ice age are 4°C–5°C (7°F–9°F). That change occurred over a period of about 7,000 years, starting 18,000 years ago. The concentration of CO_{2gas} has risen by 40% in only the past 200 years, contributing to human alteration of the planet's energy budget that has so far warmed Earth (based on only earth surface) by about 0.8°C (1.4°F). The cooling state of the molten earth took place, which led to the crust formation. In addition, it is known that the molten lava-like interior of earth is almost at constant very high (6,000°C).

>>>§Measurements of composition of air in ice cores show that for the past 800,000 years until the 20th century, the atmospheric CO_2 concentration stayed within the range 170 to 300 parts per million (ppm), making the recent rapid rise to nearly 400 ppm over 200 years particularly noticeable. During the glacial cycles of the past 800,000 years both CO_{2gas} and methane have acted as an important amplifier of the climate changes triggered by variations in the Earth and the Sun.

Furthermore, from the very warm year 1998 that later was followed by the strong 1997–98 El Niño, the increase in average surface temperature has slowed down. However, despite the slower rate of warming the 2000s were warmer than the 1990s. Decades of slow warming as well as decades of accelerated warming have occurred naturally in the climate system. Also, decades that are cold or warm compared to the long-term trend are seen in the observations of the past 150 years and also captured by climate models.

Most significantly, more than 90% of the heat added to Earth is absorbed by the oceans and penetrates only slowly into deep water. Due to the lack of mixing, the system is thus at a pseudo-equilibrium. A faster rate of heat penetration into the deeper ocean will slow the warming seen at the surface and in the atmosphere, but by itself will not change the long-term warming that will occur from a given amount of CO_2. For example, recent studies show that some heat comes out of the ocean into the atmosphere during warm El Niño events, and more heat penetrates to ocean depths in cold La Niñas. Such changes occur repeatedly over timescales of decades and longer.

This has been related to the major El Niño event in 1997–98 when the globally averaged air temperature soared to the highest level in the 20th century as the ocean lost heat to the atmosphere, mainly by evaporation. Recent studies have also pointed to a number of other small cooling in sequences over the past decade or so. These include a relatively quiet period of solar activity and a measured increase in the amount of aerosols (reactive particles) in the atmosphere due to the cumulative effects of a succession of small volcanic eruptions.

Currently, various reports are found as regards the effect of climate change on arctic (north/south). However, these are out of scope of this Appendix.

The reader is suggested to pursue about these varying conclusions as discussed in the literature (Lomborg, 2022).

The effect of winds on the sea ice is also a large factor. The changes in wind directions and in the ocean seem to be dominating the patterns of climate around Antarctica. The degree of pseudo equilibrium of carbon dioxide in air/oceans is effected as more mixing is present under storms/strong winds/etc.

>DIVERSE CARBON DIOXIDE (CO_2) SOURCES AND SINKS AND RECYCLING

The evolutionary development of the earth atmosphere over the billions of years is of current interest. However, obviously there is a lack of much knowledge about these data. In spite, one has speculated about the atmosphere development directly (Calvin, 1969).

It is useful to describe some of the major phenomena which lead to:

- production of carbon dioxide (man-made)
- production of carbon dioxide (natural sources)
- recycling of carbon dioxide
- capture of carbon dioxide (natural phenomena)(rain drops solubilization)
- capture of carbon dioxide (man-made phenomena)
- storage of carbon dioxide (man-made)
- storage of carbon dioxide (natural: in oceans/lakes/rivers + plants)
- recycling of man-made carbon dioxide

These are all related to man-made phenomena.

>PHOTOSYNTHESIS, RESPIRATION AND CO_2 ON EARTH:

Life on earth is mainly dependent on the interaction (photosynthesis) between sun, air, and water, besides other phenomena. Especially, sunlight is responsible for many life dependent phenomena. The main effect of sunlight is heat and photosynthesis. One of the most important natural phenomena on earth is photosynthesis (Rabinowitch & Govindjee, 1969). This subject is out of scope of the present context, but only a short description is given here. It is obvious that one cannot estimate accurately the total yield of photosynthesis on the earth's surface. The yield in ocean plants is very large. Some data of photosynthesis yield for different vegetation types are given in Table A.1 (Rabinovitch & Govindjee, 1969).

Figure: Sun – atmosphere (water + carbon dioxide) – land (earth) – plants

The data suggest that both the sun and earth were born simultaneously (Manuel, 2009). Sun is known to be the main source to produce heat radiation due to hydrogen fusion (fusion of hydrogen (75%) to helium (24%).

Sun consists mainly of hydrogen (H_2) 90%. The variations in solar activity is related to various parameters (e.g. cycles of solar eruptions, cosmic rays, sunspots, and variations in magnetic parity) (Manuel, 2009; IPCC, 2007; Jose, 1965; Birdi, 2020).

TABLE A.1

Typical Photosynthesis Yields of Carbon Dioxide Captured (Tons Carbon/Year) into Organic Matter

Vegetation	Area (10^6 km^2)	Tons of C/km^2	Total Yield (10^9 tons-C/year)
On Land			
Forests	44	250	11
Grassland	31	35	1.1
Farmland	27	150	4
Desert	47	5	0.22
Total on land	149	16^3	

The process of photosynthesis converts CO_2 (in conjunction with H_2O) in air to all kinds of plants and food for mankind. This process has been known to have existed on earth for millions of years.

This complicated process produces the most essential building block, glucose ($C_6H_{12}O_6$), for the formation of cellulose and carbohydrates. The main photosynthesis reaction is:

$$6\ CO_2 + 6H_2O + \text{sunlight.} \underset{\text{Catalyst}}{==========} C_6H_{12}O_6 + 6\ O_2$$

It is important to mention that photosynthesis is only observed on earth in the whole solar system. This phenomenon is known to be dependent on:

- carbon dioxide (420 ppm: 0.042)
- water
- sunshine

In this very important life sustaining reaction, the various chemicals are acquired from different sources.

- CO_{2gas} is provided by air (with a current equilibrium concentration ca. 420 ppm (042%).
- Water is present as moisture
  Or obtained from wet soil.
- The green color in plants is the chlorophyll molecule. The latter is the catalyst of the reaction. The green color in plants is related to the chlorophyll molecule.

A.6   SOURCES OF CARBON DIOXIDE (CO_2) GAS

On the earth there are currently both natural and man-made sources of CO_{2gas} (in gaseous form). The CEE indicates that carbon dioxide has existed simultaneously with photosynthesis. However, photosynthesis and GHG phenomenon have existed for many millions of years.

>LUNG FUNCTION AND HUMAN METABOLISM (CO_2 Cycle)
(PHOTOSYNTHESIS-CO_2 IN ATMOSPHERE-LIVING SPECIES):

It is reported that human metabolism and carbon dioxide (gas) play very important roles in the evolution of related species (Birdi, 2020).

Since living species are only found on planet earth, it is very useful to consider the mechanism of metabolism of these species (shortly). Especially, it is related to carbon dioxide + photosynthesis + food.

Life on earth, especially mankind, is dependent on some very essential needs; one of these is food. Food (such as: corn, rice, fruits, etc) is made by the photosynthesis of carbon dioxide. Food is used for growth through metabolism. Metabolism is dependent on the composition of atmosphere gasses (e.g. oxygen; carbon dioxide; CO_{2gas}).

The metabolism of man and many other living species is simply:

EXAMPLE:

METABOLISM
FOOD INTAKE – METABOLISM – PRODUCTION OF CARBON DIOXIDE AND OTHER WASTE PRODUCTS

Even though this description of metabolism seems very simple, it actually is the most complex phenomenon as regards the description of existence parameters of life on earth.

However, food is produced through photosynthesis (CO_2 from air). It is thus seen that the metabolism is CO_2 neutral (almost).

EXAMPLE:

Photosynthesis-carbon dioxide (CO_2)-Food Cycle

The growth of living species is dependent on the intake of food and the subsequent metabolism. Metabolism is a very complex biological-chemical aspect. In the present case:

food (as produced by Photosynthesis-carbon dioxide) intake by all living species
food metabolism and subsequent production of carbon dioxide and the necessary usage of oxygen.

The function of the respiratory system of humans (and other living species) shows that in the lungs different gases (from air) are exchanged (i.e. from blood to air in the inhaled lungs) as a result of metabolic processes. Breathing brings the oxygen (O_2) in the air ($O_2 = 28\%$) into the lungs and into close contact with the blood, which absorbs it and carries it to all parts of the body. At the same time, the blood delivers carbon dioxide (CO_2), which is carried out of the lungs when air is breathed out (Birdi, 2020).

Human metabolism:

> Blood flow into lungs $=$ Hemoglobin-CO_2
> *(Exchange of CO_2 (from air) with O_2 in lungs)*
> Blood flow out from lungs $=$ Hemoglobin-O_2

The composition of the inspired air and the expired air is found to be of following composition:

(LUNG-METABOLISM-COMPOSITION OF AIR)

	Inspired Air	Expired Air
Oxygen	21%	16%
Carbon dioxide	0.04 %	4%
Nitrogen	78.0%	78%

It is important to notice that almost equivalent amounts of increases in carbon dioxide are same as decreases in oxygen: in exhaled air breath.

EXAMPLE:

<HUMAN LUNG-CO_2-OXYGEN
    <INHALE-ATMOSPHERE

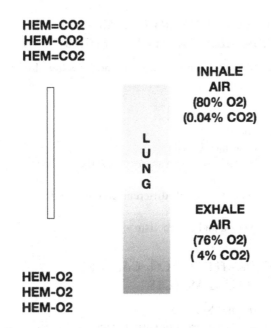

HEM=CO2
HEM-CO2
HEM=CO2

INHALE
AIR
(80% O2)
(0.04% CO2)

L
U
N
G

EXHALE
AIR
(76% O2)
(4% CO2)

HEM-O2
HEM-O2
HEM-O2

The principal function of the lung is to exchange oxygen (absorb from air) and carbon dioxide (desorb to air) between blood and inspired air. In the lungs, hemoglobin carries $CO_{2,gas}$ to the lungs and releases it by exchanging. Simultaneously, hemoglobin absorbs oxygen (21%) from air and carries to usage inside the body. Blood vessels in the lungs have a structure similar to the bronchial tree. The pulmonary artery carries de-oxygenated blood from the right ventricle to the lungs. The pulmonary artery branches first to the left and right lung and branches further down to the capillary level. The pulmonary veins carry blood from the lungs to the left atrium of the heart. They have an inverse structure compared to the pulmonary artery, starting on the capillary level and reaching the main pulmonary vein which leads to the heart.

The molecule, CO_2, is known to bind to the hemoglobin molecule, and this complex carries CO_2 to the lungs and exchanges with oxygen (from inhaled air). However, if CO_2 concentration in the air is over (30,000 ppm) 3%, then breathing becomes difficult (as may happen in closed environments, such as mines , rocket capsules, etc). In-fact, above 14% CO_2 it is fatal to human life. This shows that the food – metabolism cycle is almost neutral. Excepting, energy is used for food production, transport, etc.

Figure. : Photosynthesis-carbon dioxide – food – living species – metabolism (world population)

A.7   ROLE OF EVOLUTIONARY CHEMICAL
EQUILIBRIA (CEE) AND LIVING SPECIES

The chemical equilibrium is known to have had a profound effect on most of the evolutions:

- photosynthesis
- plants-food/fisheries
- living species and metabolism
- world pollution (air/drinking water/oceans/lakes/rivers)

For instance, the metabolism of different living species is a very significant characteristic.

<Human lung function and metabolism>

A.8   COST (ESTIMATED) OF CO_2 CAPTURE
AND RECYCLING ASPECTS

Cost of CO_2 Capture and Storage:

Although the economic aspects of carbon capture recycling and storage (CCRS) technology are out of scope of this book, a very short mention is provided. The different technologies which might be suitable for suits for CCRS will of course be dependent on the individual physical structures and background of the power plant (Birdi, 2020). As is known from other developing technologies, the cost/method are dynamic and future development may have a significant effect. Additionally, the geographic situation will also be different. One may consider the release of CO_2 in air as a pollutant. The economy will be expected to be dependent on different factors. The costs of separation may be divided into technical versus non-technical costs.

In the case of non-technical costs, one may include account depreciation and return on investment, interest rate, labor, etc. Costs associated with the technology may include the expenses related to the equipment, chemicals used, power consumption and power cost, etc. For instance, if sustainable power sources are used (such as: solar energy; wind energy), then the costs will be much different.

Additional factors that may affect the cost of CO_2 capture may also include the type of power plant and capture technology. For instance, whether one uses an existing power plant or the CCRS technology be applied to a new plant? These process design and variables are too complex and need to be investigated. Other factors which will also need to be considered are: capture capacity; capture rate; CO_{2gas} concentration (flue gas: in and out).

It is obvious that in general, the carbon capture cost will be related to the concentration of CO_{2gas} in the flue gas (Birdi, 2020). Higher the starting concentration of CO_2, lower the average cost of capture. A plot of (Figure C.1) cost increase with increasingly dilute CO_{2gas} flue gas has been reported. This kind of plot is called the Sherwood correlation (Figure A.6).

Concentration is approximately 12 mol% CO_2 (Lightfoot and Cockrem, 1999).

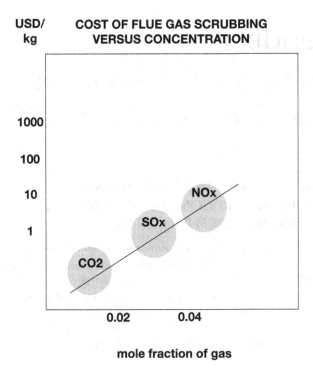

FIGURE A.6   A plot of various gas scrubbing processes' cost increase with decreasing concentration.

Appendix B
Carbon Capture Procedures – Surface Chemistry Aspects (CCRS)

B.1 INTRODUCTION: CARBON RECYCLING CAPTURE (ADSORPTION AND ABSORPTION) AND STORAGE (CRCS) PROCESSES

The earth is surrounded by gaseous atmosphere (up-to a height of 20 km). The molecular density (molecule/volume) decreases nonlinearly with height. This indicates that different properties, such as GHG, will also decrease with height. In fact, around a height of 20 km the molecular density (molecules/unit volume) is almost impossible for any living species (including mankind). This means that the GHG effect from any molecule (CO_{2gas}; CH_4; water vapor) will be completely absent around 10–20 km. In other words, the GHG effect will be varying with height, time, and place.

Earth was created about 4.6 billion years ago. It was a hot environment with unspecified atmosphere. However, it entered a cooler period, and started forming a crust. Maybe, this process is currently at a very slow pseudo equilibrium. The atmosphere appeared, and when the suitable temperature was present, some phenomenon appeared:

- photosynthesis (which needs: sunshine + carbon dioxide + water vapor + low temperature)
- living species evolved
- chemical evolutionary equilibrium (CEE) establishment
- CEE created the living species as known today

With regard to the current subject, the usage of fossil fuels is considered to add increasing amounts of carbon dioxide, CO_{2gas}, to the environment, i.e. CEE. The *average equilibrium concentration of CO_{2gas} in atmosphere* is known to be around 420 ppm (at sea level). In this context, it is useful to add that the GHG effect of CO_{2gas} has been present on earth since the photosynthesis/pre-living species/pre-fossil fuels.

Estimates indicate that $CO_{2gas} = CO_{2aq}$ alone, equilibrium over oceans, has absorbed about 50 times (Rosenzweig et al., 2021; Birdi, 2020).

The subject in this Appendix relates to controlling/mitigating/recycling the concentration of a specific substance (gas), i.e. CO_{2gas}, in the earth environment.

This phenomenon is known to be at equilibrium (Birdi, 2020; IPPC, 2011.

There are different interfaces in the sun-atmosphere-earth system (Chapter 1).

There is a need to apply surface chemistry principles of gas: adsorption and absorption, in order to concentrate the comparative low magnitude of the GHG (CO_2), which is produced from fossil fuels combustion. These processes are found to be useful in various industrial processes for effective recycling (Birdi, 2020; Gates, 2022; Rosenzweig et al., 2021). The latter processes are useful in order to mitigate effective carbon recycling in society.

EXAMPLE

>FOSSIL FUEL COMBUSTION > PRODUCTION OF CARBON DIOXIDE GAS (CO2) (10% CO2gas IN FLUE GASES)>
   RECYCLING (ADSORPTION/ABSORPTION)>
   <STORAGE-RECYCLING>

This climate – capture – recycling – storage (CCRS) phenomenon is the most important technology being used currently (Birdi, 2020; Gates, 2022).

In these phenomena, one needs to consider the molecular mechanisms where phases meet: at an interface (atmospheric – oceans; atmospheric – land). For example:

- …air (atmosphere) interacts at the interface air – oceans/lakes/rivers
- …air (atmosphere) interacts at the interface air – land (forests/deserts)

These interfacial interactions consist of varying magnitudes (specific to individual components: nitrogen; oxygen; carbon dioxide; other):

B.2   ATMOSPHERE – INTERFACE – (WATER) INTERFACE

This chapter concerns with the subject as related to:

- Gas-phase: a gaseous substance
- Solid-phase: a solid substance with surface molecules
- Gas-Solid Interface: Gas molecules interact with surfaces molecules of the solid at this interface which interact with gas molecules

SYSTEM:

Gaseous state, i.e. carbon dioxide (CO_{2gas}). As regards essential description of climate change on earth, one finds an extensive discussion on the relation to carbon dioxide (CO_{2gas}): in the gaseous state.

The purpose of this Appendix is to add some extra information, with regard to the $CO_{2,gas}$: recycling; absorption and capture aspects of CCRS (Chapters 2 and 4).

The additional information is also intended to be useful for further understanding of the present subject matter. The additional data included here will also explain some methods described in the analysis.

§CARBON ADSORPTION/ABSORPTION-RECYCLING

>Gas Adsorption on Solid Surfaces Essentials:

Solid Surfaces
Somehow, in a system with interfacial phase, one always considers that the concentration of a solute in a phase is evenly distributed. However, the Gibbs adsorption principle (Chattoraj & Birdi, 1984) shows that the solute-interfacial concentration is higher than that in the bulk phase. This may be described: for example, in systems such as:

• gas – solid system.
• liquid – solid system.

EXAMPLE:

<GAS – SOLID>
<: Variation of concentration of gas, G, at an interface of adsorbent versus distance from the adsorbate (solid (SS) or liquid (LL).>
<INITIAL>
G   G   G

SSSSSSSSSSSSSSSS

<AT GAS-ADSORPTION EQUILIBRIUM >

GGGGGGGGGGGG
SSSSSSSSSSSSSSS
SSSSSSSSSSSSSSS

AT GAD LIQUID ABSORPTION

GGGGGGGGGGG
LLLLLLLLLLLLL
LLLLLLLLLLLLL

For example, a gas such as carbon dioxide (CO_{2gas}) in air is reported to be about 0.042% (at the sea level; ambient temperature: STP). Current reports in the literature consider this quantity being high. Further, the high concentration of CO_{2gas} has been correlated with the climate change, due to its GHG (Chapter 1) (Birdi, 2020; Rosenzweig et al., 2021).

Besides, carbon dioxide in air/oceans is the sole source of carbonaceous substances as found in nature. Thus carbon dioxide, CO_{2gas}, has existed throughout the photosynthesis-living species ages.

In addition, the CEE phenomenon has indicated this conjecture.

However, this quantity is only valid for a particular place on the earth and serves as a reference point in space and time. This magnitude is an estimate quantity and varying with time and place.

This magnitude is known to vary with long or short time. The variability has to be based on geological scale of time/evolution.

Furthermore, one also considers that the concentration of $CO_{2,gas}$ is evenly distributed. This is of course correct, though within certain constraints. Due to a larger molecular weight of $CO_{2,gas}$ (44 g/mol) than other molecules in air, the latter is found to be unevenly distributed in atmosphere (i.e. its concentration varies with height, earth land/oceans/lakes/rivers, forests). There exists a large nonlinear character.

In other words, the concentration of $CO_{2,gas}$ in atmosphere will be nonlinear in distribution (as regards: place, height, and summer/winter).

In the present case, the state of CO_{2gas} in atmosphere around the earth will be considered.

EXAMPLE

System:
  PHASE-I: Air (Atmosphere)
  PHASE-II: Oceans (+lakes/rivers)
  >INTERFACE: Between Phase-I/ Phase-II
      >THIS INTERFACE COVERS 70% OF EARTH
      >MAJOR QUANTITY OF HEAT BALANCE OCCURS AROUND THIS
  LARGE INTERFACE

This interface creates another special system. There exists an (pseudo) equilibrium between the concentrations of $CO_{2atmosphere}$ (CO_{2gas}) and $CO_{2,oceans}$ (CO_{2aq}).

Since evolutionary processes have exhibited that these have attained an equilibrium, one might also expect similar equilibrium in the case of (i.e. CEE):

$$(CO_{2gas}/)/(CO_{2aq}) == K_{co2}$$

(where aq – oceans-lakes-rivers; == indicates equilibrium state)

The atmosphere surrounding the earth is rather complex (Chapter 1; Appendix A). The gas molecule in atmosphere interacts with the earth:

At interfaces:

<atmosphere – (30%) land (plants/forests/deserts/Antarctica)

-(70%) oceans (with depths more than 5 km)

The gas molecules in atmosphere are attracted to the earth (gravity forces) with decreasing force as one moves away from the earth. This is experienced by all while going up in a plane or rocket. The need for oxygen at certain height (about 5,000 m) is well known. In fact, at heights around 10 km, humans need oxygen for normal breathing. This arises from the fact that:

- molecular density is very low at these heights
- the magnitude of mean-free-path increases to over few km.
- atmospheric pressure gets low
- lung function (humanoids) gets reduced for proper breathing (lethal).

EXAMPLE

STATE OF CO_2-GAS IN DIFFERENT PHASES

>IN ATMOSPHERE

Content of CO_{2gas} in air = 0.042% (420 ppm)

CO_{2gas} is soluble in water (at STP)

>AIR/WATER INTERFACE:

Air near water surface leads to the solution of CO_{2gas} (denoted as: CO_{2aq})

This means that the first layer (few molecular layers) of the ocean water is containing $CO_{2g,aq}$ from CO_{2gas}.

CO_{2gas}......................in air

CO_{2aq}.................... surface of water

This means that the concentration of CO_{2gas} near the surface of water (especially in terms of molecular dimensions) is very low due to its solubility in water (formation of CO_{2aq}). However, one finds the phenomena of diffusion of CO_{2aq} in oceans (with depths varying up to over 5 km (5,000 m = ca. 5,000 yards) (Appendix A). Further, the temperature (ca. 100°C) and pressure (ca. 500 atm (= bar) at these depths can be significant (Birdi, 2020).

B.3 ADSORPTION OF GAS ON SOLID SURFACE

The process of adsorption of a gas molecule on a solid surface has been investigated for more than about a century (Chattoraj & Birdi, 1984; Adamson & Gast, 1997; Myers, 1997 Birdi, 2002, 2016; Tovbin, 2017; Birdi, 2020).

Molecules in the gas are moving larger distances (ten times), as compared to the molecules in liquid or solid state (Birdi, 2020). The volume per molecules in the gas phase is about 1,000 times larger than that in liquid or solid. Additionally, the gas density in atmosphere decreases nonlinearly with increasing height. This means that the number of molecules per unit decreases with height. These observations indicate that the GHG effect decreases with height.

Under suitable conditions, gas molecules interact with a solid surface in different ways, as regards the surface forces. When a gas molecule comes close to a solid surface or a liquid surface, the following may take place (Chapter 2):

- Gas molecules may bounce back into the gas phase (absence of adsorption);
- or adsorb at the surface (under the field of force of the solid surface atoms).
- Gas may also exhibit preference for adsorption on a specific solid site.

In the case of a solution (solid substance dissolved in a liquid), adsorption is the phenomenon marked by an increase in density of a fluid near the surface, for our purposes, of a solid. In the case of gas adsorption, this happens when molecules of the gas occasion to the vicinity of the surface and undergo an interaction with it, temporarily departing from the gas phase. Molecules in this new condensed phase formed at the surface remain for a period of time, and then return to the gas phase. The duration of this stay depends on the nature of the adsorbing surface and the adsorptive gas, the number of gas molecules that strike the surface and their kinetic energy (or collectively, their temperature), and other factors (such as capillary forces and surface heterogeneities). Adsorption is by nature a surface phenomenon, governed by the unique properties of bulk materials that exist only at the surface.

A gas molecule/atom (adsorbent) can interact with a solid (surface) with varying types of interaction forces or mechanisms. These are mentioned as:

- chemical adsorption (chemisorption)
- weak physical adsorption (physisorption)
- penetration into pores (porous solids) (absorption).

Adsorption of a gas molecule (from gas phase) thus gives rise to a lowering in entropy on adsorption (Adamson & Gast, 1997; Keller & Stuardt, 2006; Bolis et al., 1989; Birdi, 2009, 2020).

In the current literature, one finds that the description of thermodynamics of physical adsorption of gases in porous solids has been investigated (Chapter 2). The measurement of the amount of gas adsorbed by a solid is carried out by volumetric or and gravimetric techniques based upon the concept of the (Gibbs) surface excess model (Chattoraj & Birdi, 1984; Myers, 1989).

The subject of gas adsorption on solids has been investigated by principles of surface thermodynamics (Gibbs adsorption theory) (Chattoraj & Birdi, 1984).

Basically, surface thermodynamics analysis provides quantitative relationships between phenomena such as the amount adsorbed and the heat and entropy of gas adsorption. However, in the case of adsorption in porous solids, the surface tension (of the solid) (Chapters 2 and 3) and surface area are not easily available by any direct method. However, in the case of microporous solids, the gas adsorption phenomena will be different. The gas molecules which are adsorbed inside the pores (of varying size) will interact differently than those adsorbed in large pores.

B.4 SOLID MATERIALS USED FOR GAS ADSORPTION

Various solids have been used for gas adsorption processes. Accordingly, these processes are delineated in this section (Birdi, 2020).

B.4.1 DIFFERENT SOLID MATERIALS USED FOR GAS ADSORPTION

In the present case, the main interest is as regards the state of carbon dioxide gas molecules.

The carbon (carbon dioxide gas) capture process is based on using different kinds of solids. The primary objective is to control/analysis/capture/store/recycling of the man-made increase in carbon dioxide (CO_{2gas}) production (by industrial/domestic) fossil fuel comb. The characteristics of the solid have to be investigated, in order to find the most appropriate for a given process. The gas – solid adsorption process needs to be determined suitable for each system under consideration.

The different solids used for gas adsorption are described here.

B.4.2 POROUS SOLID MATERIALS (USED FOR GAS-ADSORPTION)

The surface properties of a solid are dependent on various factors. The most important one arises from the size of solid particles. Finely divided solids possess not only a geometrical surface, as defined by the different planes exposed by the solid, but also an internal surface due to the primary particle aggregation. This leads to pores of different sizes according to both the nature of the solid and origin of the surface. Experiments show that these pores may be circular; square, or other shape. The porous solids:

/S/ //S/ //S/ /S/ //S/ /S/ //S

(where: S = solid surface; / // = pores)

The size of pores is designated as the average value of the width, w (Birdi, 2017, 2020). The width, w, gives either the diameter of a cylindrical pore, or the distance between the sides of a slit-shaped pore.

The smallest pores, with the range of width $w < 20$ Å (2 nm = 2 10^{-9} m = 20 10^{-8} cm = 20 Å), are called *micropores*.

The *mesopores* are in the range of a width 20 Å $\leq w \leq$ 500 Å (500 10^{-8} cm) (2 and 50 nm).

The largest pores, in the range of width $w > 500$ Å (50 nm), are called *macropores* (Birdi, 2017). The shapes of pores will vary in geometric size and shape (e.g. circular, square, and triangular). The capillary forces in these pores will thus depend both on the diameter and on the shape.

In general, most solid adsorbents exhibit (for example: like charcoal and silico-alumina) irregular pores with widely variable diameters in a normal shape. Some other adsorbents, conversely, other materials such as zeolites and clay minerals are entirely micro- or meso-porous, respectively. In other words, the porosity in these materials is due to the primary particle aggregation but this is an intrinsic structural property of the solid material (Birdi, 2017, 2020).

B.4.3 DIFFERENT SOLID ADSORBENTS USED FOR CO_2 (GAS) CAPTURE (ADSORPTION)

The aim of adsorption of a gas on solid is primarily to obtain almost 100% pure gas, in the process. For instance:

- \>carbon dioxide in air = 420 ppm (0.042%)
- \>carbon dioxide in flue gases = 10,000 ppm; (10% (Birdi, 2020). Therefore, it is useful to mention a few examples here for the sake of description of the mechanisms of the process. Obviously, there is a specific requirement for a suitable absorbent of CO_2. (Chapter 4). One of these requires that the adsorption – desorption characteristics of gas should be acceptable for the process. For instance, the adsorbed CO_{2gas} (under high pressure) could be desorbed on change of pressure (reduction of pressure) or stripping with a suitable gas, such as stream with an inert gas (or some similar process). Further, from surface chemistry principles, the adsorbent should exhibit a large surface area per weight. The physical adsorption process has many advantages over other methods. It is of a low energy system. The rate of adsorption – desorption step is comparatively short. These different phenomena will thus need to be adapted to the specific application.

§ZEOLITES USED FOR GAS ADSORPTION: Gas adsorption on solids, such as **zeolites**, has been investigated. Zeolites are available either as natural state or as synthetic crystalline alumino-silicates, the structure of which is based upon a three dimensional polymeric framework, with nanosized cages and channels (Breck, 1974; Birdi, 2020).

The basic building block of such materials, of general formula: $Mn + (AlO) (SiO)^{x-}$ zHO, is the [TO] unit with T = Si, Al.

These studies showed that this unit is a tetrahedron centered (as determined from X-ray analyses) on one T atom bound to four O atoms located at the corners; each O atom is in turn shared between two T atoms. Further, these tetrahedral units join each another through T–O–T linkages in a variety of open-structure frameworks characterized by (interconnected) channels and voids which are occupied by cations and water molecules (Lowell et al., 2006; Rabo, et al., 1981; Yang, 1987).

The presence of charge-balancing (extra-framework) of cations is required in order to compensate the negative charge of the tetrahedral $[AlO_4]^-$ units in which Al is in isomorphous substitution of Si atoms. The density of charge-balancing cations depends upon the Si:Al ratio (i.e. in the range from 1 to ∞).

The most significant adsorption characteristics of zeolites arise from the presence of the nano-sized *cages and channels* within the crystalline structure. Further, the shape of channels (which gives rise to selectivity properties) makes these useful solid materials for different applications: e.g. catalysis and gas separation processes. The most characteristic property of any porous materials is that they exhibit high surface areas (i.e. area/gram of solid: over $3,000\,m^2/gm$), which maximize the extension of the interface region.

Furthermore, in both imperfect or perfect all-silica zeolites, the process was entirely reversible upon evacuation of the gas phase. In this latter case, NH_3 interacted only *via* hydrogen bonds with $Si-OH$ nests. In the defect-free MFI–silicate, which exposes only non-reactive siloxane bridges, the interaction was a specific which was governed by dispersion forces due to the nano-porous walls (*confinement effect*).

The process of gas adsorption is dependent on both the characteristics of the gas (absorbate) and the solid (adsorbents) (besides temperature and pressure). The surface characteristic of the solid needs to be of suitable for the gas adsorption process.

<Adsorbent Solid Carbon (C_{solid}):

In the case of solid C_{solid}, it has been found in many different kinds of forms. For example, carbon black (C_{black}) is one of the forms of carbon which is produced by the incomplete combustion of heavy petroleum products such as FCC (fluid catalytic cracking) tar, coal tar, ethylene cracking tar, and vegetable oil. C_{black} is a form of amorphous carbon that has a high surface-area-to-volume ratio. However, in spite of that fact, carbon black, due to specific conductivity and mechanical properties, is not being used as a sensing material in gas sensors. Only activated carbon, also called activated charcoal, activated coal or carbon activates, one can find in gas sensors where C_{black} can be used as a filter. Activated carbon is a form of carbon that has been processed to make it extremely porous and thus to have a very large surface area available for either adsorption or chemical reactions. Due to its high degree of microporosity: one finds that 1 g of activated carbon has a surface area in excess of $2,000\,m^2$. It is useful to compare this to talcum: with a known low magnitude of solid surface area: $10\,m^2/g$

The most significant carbon black (C_{black}) surface properties useful for composite design are:

- dispersion,
- stability of the carbon black structure or network,
- consistent particle size,
- specific resistance, structure, and high purity.

The solid Carbon black (C_{black}) is used mainly in polymer-based composites. The carbon black, C_{black}, endows electrical conductivity to the films, whereas the different organic polymers such as poly(vinyl acetate) (PVAc), polyethylene (PE), poly(ethylene-*co*-vinyl acetate) (PEVA), and poly(4-vinylphenol) (PVP) are sources of chemical diversity between elements in the sensor array. In addition, polymers function as the insulating phase of the carbon black composites. The concentration of CB in composites is varied within the range of 2–40 wt%. The conductivity of these materials and their response to compression or expansion can be explained using percolation theory (McLachlan et al., 1990).

The compression of a composite prepared by mixing conducting and insulating particles leads to increased conductivity, and, conversely, expansion leads to decreased conductivity.

Micropores, where most adsorption takes place, are in the form of two-dimensional spaces between two graphite-like walls, two-dimensional crystallite planes composed of carbon atoms. The distance between the two neighboring planes of graphite has been determined to be 3.76 Å (3.76 10–8 cm: 0.376 nm).

Most activated carbons contain some oxygen complexes (traces) which arise from either source materials or from chemical adsorption of air (oxidation) during the activation stage or during storage after activation.

Furthermore, activated carbon (Cact) also contains to some extent ashes derived from starting materials. The amount of ash varies from 1% to 12%. Ashes are known to consist of silica, alumina, iron, and alkaline and alkaline earth metals. These ashes exhibit some specific characteristics as follows:

1. an increase in hydrophilicity of activated carbon,
2. catalytic effects of alkaline, alkaline earth, and some other metals such as iron during activation.

One of the most important solids is carbon (of different forms). In general, one finds that activated carbons in commercial use are present in two forms:

- powder form
- and granular or pelletized form.

Decolorization in different refinery processes, removal of organic substances, odor, and trace pollutants in drinking water treatment, and wastewater treatment are the main applications of liquid phase adsorption.

Carbon molecular sieves: The size of micropores of the activated carbon, C_{active} is determined during pyrolyzing and activation treatments. Hence, small and defined micropores that have molecular sieving effects can be prepared by using proper starting materials. The main applications of activated carbon with molecular sieving ability have been as follows: the separation of nitrogen and oxygen in air on the basis of difference of diffusion rates of these gases in small micropores; control of fragrance of winery products where only small molecules are removed.

Activated carbon fiber: Another type of carbon solid is the synthetic fibers such as phenolic resin (Kynol R), polyacrylic resin (PAN), and viscose rayon. Most ACFs have fiber diameters of 7 pm (pm= m) to 15pm, which is found to be even smaller than powdered activated carbon.

<Gas – Solid Adsorption Enthalpy (Calorimetric Methods):

The quantity most characteristic of any reaction relates to its enthalpy. The latter relates to whether the system:

- enthalpy produces heat (exothermic0)
- enthalpy uses heat (endothermic)
- enthalpy is zero (i.e. no bonds are broken/formed)

The gas adsorption on a solid can be studied by various suitable apparatus (calorimeters) as regards the change in temperature. There are many calorimeters commercially available which can provide this information. The principle of these procedures is to measure the heats of gas adsorption on solids (with varying sensitivity; accuracy), derived from direct calorimetric methods which are based on the measurement of the heat evolved when a known amount of gas is allowed to adsorb onto a "clean" surface. A "clean" surface is a solid surface kept in high vacuum conditions after having been activated either in vacuo in order to eliminate (either totally or partially) the surface contaminants, or in controlled atmosphere/conditions.

The gas solid experiment is carried out by measuring the change in temperature of the solid. Further, it is worth recalling that not only the magnitude of the heat evolved during adsorption but also its variation upon increasing coverage may reveal useful information concerning the type of adsorbate/surface sites bonding, and its evolution according to the surface heterogeneity.

EXAMPLE

>HEAT OF GAS ADSORPTION ON SOLIDS:

....

It is well known (Chapter 3) that the surface of a real solid material is, in general, characterized by a structural and/or a chemical heterogeneity of the sites, owing to the presence of either structural defects and/or (hetero) atoms in different oxidation states. Another kind of surface heterogeneity, originated by the presence of lateral interactions among adsorbed species, is the so-called induced heterogeneity.

The quantity related to the heat measured (calorimetric cells) represents the enthalpy change associated with the adsorption. This result applies to adsorption processes performed in a *gas-solid open* system through the admission of the adsorptive on the solid material kept isothermally within a heat-flow micro-calorimeter consisting of two cells in opposition.

The thermodynamics of gas – solid adsorption has been investigated by using a calorimeter (Adamson & Gast, 1997: Kellar, et al., 1992; Lowell et al., 2006; Bolis et al., 1990; Auroux, 2013).

HEAT OF ADSORPTION FOR GAS ON SOLIDS: MICROCALORIMETRY: There are commercial apparatus available for measuring gas adsorption on solids. The stepwise adsorption microcalorimetry technique is useful for providing quantitative data in surface chemistry studies. These data provide information about the nature of the adsorption process: physical or chemical adsorption. The solid surfaces have been extensively investigated by various spectroscopic techniques (Infra-red (IR) and Raman, UV-vis, NMR, XPS, and EXAFS-XANES) (Ertl, 2003).

B.4.4 DIVERSE PROCESSES IF GAS ADSORPTION ON SOLIDS: ESSENTIAL PRINCIPLES

Gas adsorption on solids: The thermodynamic adsorption principles have been investigated by various methods: such as calorimetry.

The heat (enthalpy) of gas adsorption on solids has been measured by using different types of calorimeters (as related to size/sensitivity/application). The heat evolution process when a gas or a fluid interacts with the solid surface is related to the nature and energy of the adsorbed species/surface atom interactions. The surface forces determine the thermodynamic process (exothermic or endothermic reaction). The enthalpy may be positive or negative. This provides a very useful information as regards the mechanism of the property. Further, the knowledge of the energetics of chemical and physical events responsible for the process as well as the assessment of the associated thermodynamic parameters contributes to a molecular understanding of phenomena taking place at any kinds of interfaces (Auroux, 2013; Chattoraj & Birdi, 1984; Bolis et al., 1990; Birdi, 2002, 2020).

In some cases, these studies have been correlated with other investigations by spectroscopy – atomic microscopic studies.

The quantity enthalpy (q_{ads}) measured is analyzed as follows. The entropy ΔS_{ads} and enthalpy ΔH_{ads} are estimated as follows.

$$\Delta S_{ads} = q_{ads} + R - RT\ln\left(p^{1/2}\right) \tag{B.1}$$

The expression on the right-hand side of the formula, the enthalpic and the pressure contributions to the standard entropy of adsorption are given. The quantity enthalpy is expressed as:

$$\Delta H_{ads} = q_{ads} / T + R \tag{B.2}$$

which is obtained from the calorimetric data, whereas the free energy term:

$$\Delta G_{ads} = -(R\,T)\ln\left(p_{gas}^{1/2}\right). \tag{B.3}$$

is obtained from the adsorption isotherms. The quantity $p_{gas}^{1/2}$ is the equilibrium pressure when half of the surface is covered with gas.

For example: The data of CO adsorption on Na–MFI and K–MFI were analyzed (at $T = 673\,K = 400\,C$). The calorimetric heat of adsorption was ca. 35 and 28 kJ mol^{-1} for Na–MFI and K–MFI, respectively. The half-coverage equilibrium pressure (obtained by the adsorption isotherms) was $p^{1/2} = 200$ Torr for Na–MFI and 850 Torr for K–MFI. The magnitude of standard adsorption entropy was

$$\Delta S_a^o = -151\ J\ mol^{-1}K^{-1}\ (for\ Na - MFI$$

$$and\ -140\ J\ mol^{-1}K^{-1} for\ K - MFI)$$

<CO adsorption data> from these studies, it was found that the decrease of entropy for CO adsorbed at a polar surface through electrostatic forces was slightly larger than that for the specific interaction of Ar atoms adsorbed at a polar surface. This is as one would expect from physical interactions. The magnitude of entropy of adsorption, S_{ads}, of a gas will be expected to be related to the energy of adsorption. This was found from the data of adsorption of CO on MFI. The change (loss) of entropy for CO adsorbed on Na–MFI was found to be larger than that on K–MFI. This was in accord with the higher energy of adsorption of CO on Na+ than that on K+ sites.

There are also reported gas adsorption studies where more than one kind of adsorbed species may be present. This was found in the case of CO adsorption on TiO$_2$. In the case of the adsorbent, it was found that in the TiO$_2$ (dehydrated) surface, CO adsorption showed two ad-species.

This was ascribed to the existence of two distinct IR bands located at $v_{CO} = 2,188$ and $2,206$ cm^{-1} (Bolis et al., 1989). These analyses showed that there were present two ad-species which were formed on two different Lewis acidic sites made up of structurally different *cus* Ti^{4+} cations (species A and B). These A and B species showed the following spectroscopic and energetic properties:

- Species A ($v_{CO} = 2,188$ cm^{-1}) were formed at the 5th-coord Ti^{4+} cations typically exposed at the flat planes of anatase nanocrystals;
- species B ($v_{CO} = 2,206$ cm^{-1}) were formed at the 4-coord Ti^{4+} cations, which are exposed at the steps, corners, and kinks of the flat planes (Rouquerol et al., 1998). It was found that species – A v_{CO} frequency moved from 2,188 down to 2,184 cm^{-1}.

<DIFFERENT CALORIMETERS USED FOR GAS ADSORPTION:

MICRO-CALORIMETER: different – Calvet heat-flow microcalorimeters are an example of high sensitivity apparatus which are suitably adapted to the study of gas-solid interactions when connected to sensitive volumetric systems. Volumetric-calorimetric data reported in the following were measured by means of a commercially available, standard heat-flow microcalorimeter.

In these studies, both the integral heats and adsorbed amounts were measured. Two identical calorimetric vessels, one containing the sample under investigation, and the other (usually empty) serving as the reference element, were connected in opposition.

In some studies, a stepwise procedure was used (Bolis et al., 1999). Small successive doses of the adsorptive were admitted and left in contact with the adsorbent until the thermal equilibrium was observed. After a definite amount of gas was introduced into the calorimeter, the amount of heat, ΔQ_{int}, was measured, while the adsorbed amount of gas, Δn_{ads}, was measured separately. Integral heats normalized to the adsorbed amounts are referred to as the integral molar heat of adsorption at the given equilibrium pressure p_{gas}:

$$\left(q_{mol}\right)p_{gas} = \left(Q_{int}/n_{ads}\right)p_{gas} \tag{B.4}$$

in kJ mol^{-1}. The quantity q_{mol} refers to an intrinsically average value, and is related to different thermal contributions from the interactions between the gas molecules. In the case of non-interacting binding sites, the heat of adsorption is constant, regardless of the degree of surface coverage. Hence, the integral heats curve is thus a straight line through the origin, and the slope is equal to the differential heat of adsorption (q_{dif}). It is necessary to understand the gas adsorption process, as regards the dependence of degree of coverage and molecular interactions. This information is obtained from heats of adsorption data as a function of the degree of adsorption. The magnitude of the heat evolved during adsorption, which depends on the nature of the adsorbate/surface sites bonding, varies upon increasing coverage as a consequence of the presence of either a heterogeneous distribution of surface sites, or lateral interactions among adsorbed species.

For example: The adsorption of water, H_2O, on H−BEA and BEA-zeolites has been analyzed.

These studies were carried out as a function of water adsorbed amounts or water equilibrium pressure.

From these studies it was concluded that it will be reasonable to expect that at the Si(OH)$^+$Al$^-$ sites water (H_2O) molecules are adsorbed (with strong hydrogen-bonds).

The heats of adsorption started from a quite high zero-coverage value ($h_{ad} \approx 160$ kJ mol^{-1}), which is compatible with a chemisorption process, either the protonation of H_2O at the (Bronsted) acidic sites or the strong oxygen-down coordination at the (Lewis) acidic sites. For each H_2O molecule adsorbed per Al atom, on average, the heat values were found to be in the range of $160 < h_{diff} < 80$ kJ mol^{-1} range, whereas for the second-to-fourth H_2O adsorbed molecules in the $80 < h_{diff} < 60$ kJ mol^{-1} range.

In the all-silica BEA specimen, the zero-coverage heats of adsorption were much lower than those for H−BEA ($h_{ad} \approx 70$ vs.160 kJ mol^{-1}, respectively).

These data thus suggested that the all-silica zeolite behaves as a non-hydrophobic surface.

A constant value for the differential heat was obtained: $h_{dif} \approx 35$ kJ mol^{-1} for Na−MFI and ≈ 28 kJ mol^{-1} for K−MFI. The linear fit of the integral heat curves seemed the most realistic, in spite of the fact that in both cases at very low and at high coverage the middle points of the experimental histogram deviated from the constant value. In fact, the low-coverage heterogeneity was due to the presence of a few defective centers (1%–2% of the total active sites) interacting with CO more strongly than the alkaline metal cations.

For example: The enthalpy of adsorption of ammonia (NH_3) has been reported in the literature (Bolis et al., 1989).

NH_3 adsorption on H −MFI and all-silica MFI zeolites: The data for enthalpy of the reversible adsorption of NH_3 on different MFI−Silicalites (Sil−A, Sil−B, and Sil−C) and on a perfect (i.e. defect-free) MFI−Silicalite (Sil−D) have been studied. The values of heats of adsorption started from a quite high zero-coverage value ($h_{ad} \approx 160$ kJ mol^{-1}), which is compatible with a chemisorption process, either the protonation of H_2O at the Brønsted acidic site or the strong oxygen-down coordination at the Lewis acidic sites.

These studies showed that one H_2O molecule adsorbed per Al atom, on average, the heat values were found to be in the range of $160 < h_{dif} < 80$ kJ mol^{-1} range, whereas for the second-to-fourth H_2O adsorbed molecules in the $80 < h_{dif} < 60$ kJ mol^{-1} range.

It has been explained that the volumetric technique is more accurate at low pressure because almost all of the metered doses are adsorbed. The gravimetric technique has the disadvantage at low pressure that the amount adsorbed is the difference of two nearly equal numbers. At high pressure, the volumetric technique gives the amount adsorbed as the sum of a large number of doses with an associated cumulative error. The gravimetric technique is more accurate at high pressure because the measured amount adsorbed is referenced to the weight of the adsorbent in a vacuum.

Volumetric method: The volumetric technique is to introduce a known mass of adsorbent into a sample cell of calibrated volume.

B.4.5 Additional Methods of Gas Adsorption on Solids

In the literature one finds various methods which have been used to study the process of gas adsorption on solid surfaces. This process has been studied for almost a century. Further, it is also one of the most developing processes. This arises from the fact that the gas adsorption plays a very important role in many important industries (such as gas purification and separation technology; catalysis; flue gas treatment; water purification; pollution control; gas sensor technology).

The different methods used depend on the system under investigation. There are also many commercially available instruments which are designed for any specific process under investigation.

> **Gravimetric method**. In the gravimetric method the amount of substance adsorbed on another phase is measured. A mass of solid adsorbent is loaded into a container attached to a microbalance. Following desorption of the solid using high temperature and vacuum, the system is brought to a specified temperature and gas is admitted to the sample cell. After adsorption is complete, at equilibrium, the temperature (T) and pressure (P) are measured and the adsorption is determined from the weight of the solid + adsorbed gas. The weight of gas adsorbed is equal to the weight of the container with the solid minus its degassed tare weight under vacuum.
>
> **Volumetric method**: In this procedure the change in volume of the system is measured. The change in volume corresponds to the gas adsorbed (Adamson & Gast, 1997; Keller & Stuart, 2006; Birdi, 2001).
>
> **Porous solid adsorbents**: In the case of porous solids, the gas adsorption needs a different approach.
>
> The pore volume of the solid (vp) has been estimated from the amount of an inert gas (helium) adsorption in the pores at ambient temperature (Birdi, 2020).

As explained above, every solid exhibits a specific surface property: area/gram.

This is the quantity which is essential in the analyses of all gas adsorption data on solids. A general procedure is used to determine the adsorption of an inert gas, such as Helium, on the solid (Myers, 1989; Adamson & Gast, 1997; Birdi, 2020).

>OTHER GAS-SOLID PHENOMENA (SENSORS)

GAS SENSOR ON SOLID MATERIALS:

Any adsorption process is known to affect the surface-chemical properties of a solid surface.

Any specific gas adsorption process on a solid gives rise to a change in the physical properties of the solid. The latter observation thus leads to the possibility of using the change in solid characteristics as a gas sensor (Korotcenkov, 2014).

Different measuring techniques show that these changes in the solid properties are related to the gas adsorption. Hence, this observation can be used as a sensor for the gas.

It is found that it is not easy to characterize an ideal sensing materials. One finds a whole range of gas sensors in industrial applications. This shows the important application of the gas – solid adsorption process. For instance, one finds sensors which can detect a variety of gases: CO_2, CO, H_2, NO_2, NH_3, Cl_2, and H_2 (Korotcenkov, 2014).

This observation shows the following factors:

- ...gas adsorption on solid
- ...creation of a new solid surface with different physical properties.

CO_2 Capture from air:

There are literature investigations which analyze the technology which could capture CO_2 from air (400 ppm). It is thus useful, since it can be applied at all suitable sites worldwide. Of course, the main challenge for the efficient process is that the concentration of CO_{2gas} is relatively low (Dubey et al., 2002; Gates, 2021).

Cryogenic distillation of CO_2:

In any CCRS process, the aim is to produce almost pure CO_2. Mainly, this procedure leads to either usage of CO_2 in industry as well as storage in geological reservoir.

There exists another method by which CO_2 can be extracted from the flue gas. This is based on the cryogenic distillation technology.

This process consists of where a gas is separated from flue gas, by using distillation at very low temperature and high pressure, which is similar to other conventional distillation processes except t it is used to separate components of the gaseous mixture (due to their different boiling poi instead of liquid. For CO_2 separation, flue gas containing CO_2 is cooled to de-sublimate temperature (−100°C to −135°C) and then solidified CO_2 is separated from other tight and compressed to a high pressure of 100–200 atmospheric pressure. The amount of CO_2 recovered can reach 90%–95% of the flue gas. Since the distillation is conducted at extremely low temperature and high pressure, it shows an energy intensive process estimated to be 600–660 kWh per ton of CO_2 recovered in liquid form.

CO_2 Hydrate Formation and CCRS:

It is found that ice has a peculiar and abnormal characteristic and is reported to exhibit the inclusion property.

This procedure is based on a specific property of ice and hydrate formation (Sloan 1998; Birdi, 2020).

In general, all substances in the solid state have higher density (with few exceptions: water) than those in the liquid state. Ice (icebergs) floats on water, which is an exception (anomalous behavior). Ice is about 10% lighter than liquid water, at about 0°C (Birdi, 2020).

It is known that certain gas molecules form hydrates with ice (CH_4; CO_2; Cl_2) (Tanford, 1980; Birdi, 2016). Another new procedure is reported for CCRS. This is based on the formation of the complex (hydrate) between water molecules in ice and a hydrate-gas molecule. A threefold dimension model of chlorine hydrate of composition is formed (Birdi, 2016, 2020).

<CO_2-HYDRATE: It has been reported that a complex of clathrate 1:8, CO_2: 8 H_2O is formed.

It has been suggested that this structure formed cannot be used for transport. A procedure is similar to that of natural gas (CH4:WATER) clathrate.

B.5   PHYSICAL PROPERTIES OF CARBON DIOXIDE (CO_2)

A short description of some important chemical properties is useful at this stage. The different chemical characteristics are delineated under separate section. This is useful in understanding the different processes as mentioned in the text.

CO_2 is a gas at standard temperature and pressure (1 atm). CO_2 is only found in the earths' atmosphere (currently ca. 410 ppm). There has been observed an increase in CO_2 concentration in the atmosphere over the last few decades. In the molecule of CO_2, the distance between C and O is 116.3 ppm (which corresponds with double bond). Hence, the CO_2 molecule is planar (O=C=O) and linear in space. The molecule has no electrical dipole. Some physical properties of CO_2 are given in the following.

<Physical properties of CO_2>

Molecular weight	**44.03 g/mole**
Color	Colorless and odorless
Density/temperature	
Pressure	1.562 g/mL/Solid/1 atm/−78.5°C
	0.77 g/mL/Liquid/56 atm/20°C
	1.977 g/L/Gas/1 atm/0°C
Melting point	−78°C (194.7 K) (sublimation)
Boiling point	−57°C (216.6 KJ/t.185 bar)
Solubility in water	1.45 g/L at 25°C, 100 kPa (1 atm) (Carbon dioxide in air is found to interact with water (in olr) and form H_2CO_3/Carbonates/shells/fisheries)
Refractive index (nD)	1.1120
Viscosity	0.07 cP at −78°C
Molecular shape (O=C=O)	Linear
Dipole moment	zero
Infra-Red (IR) Absorption	CO_2 absorbs in IR-(>1,000 nm)
Toxicity of CO_{2gas} (to mankind/mammals)	>1% (10,000 ppm) in air (see: Appendix A/B)

References

Adamson, A. W., and Gast, A. P., *Physical Chemistry of Surfaces*, 6th ed., Wiley Interscience, New York, 1997.

Auroux, A. (ed.), *Calorimetry and Thermal Methods in Catalysis*, 3 Springer Series in Materials Science 154, Springer-Verlag, Berlin, Heidelberg, 2013.

Birdi, K. S. (ed.), *Handbook of Surface & Colloid Chemistry*, 2nd ed., 2002.

Birdi, K. S. (ed.), *Handbook of Surface & Colloid Chemistry*, 3rd ed., CRC Press, Boca Raton, FL, 2009.

Birdi, K. S. (ed.), *Handbook of Surface & Colloid Chemistry*, 4th ed., CRC Press, Boca Raton, FL, 2016.

Birdi, K. S., *Scanning Probe Microscopes (SPM)*, CRC Press, Boca Raton, FL, 2003.

Birdi, K. S., *Self-Assembly Monolayer (SAM) Structures*, Plenum Press, New York, 1999.

Birdi, K. S., *Surface Chemistry of Geochemistry of Hydraulic Fracturing*, CRC Press, Boca RRATON, U.S.A., 2017.

Birdi, K. S., *Surface Chemistry of Carbon Capture*, CRC Press, Boca Raton, FL, 2020.

Birdi, K. S., *Trans. Faraday Soc.*, 17, 194, 1982.

Birdi, K.S., and Vu, D.T., *Adhes. Sci. Technol.*, 7, 485, 1993.

Birdi, K. S., Vu, D. T., and Winter, A., *J. Phys. Chem.*, 93, 3702, 1989.

Bolis, V., Morterra, C., Volante, M., Orio, L., & Fubini, B., *Langmuir*, 6, 695, 1990.

Bryngelsson, M., and Westermark, M., *Proceedings of the 18th International Conference on Efficiency, Cost, Optimization, Simulation and Environmental Impact of Energy Systems*, 703, 2005

Calvin, M., *Chemical Evolution*, Clarendon Press, Oxford, UK, 1969.

Chattoraj, D., and Birdi, K. S., *Adsorption and the Gibbs Surface Excess*, Plenum Press, New York, 1984.

Clifton, T., *Gravity*, Oxford Press, Oxford, UK, 2017.

Cox, M. D., *Numerical Methods of Oceans Circulating*, National Academy of Science, Washington, DC, 1975.

Dubey, M. E., Ziock, H., Rueff, G., Elliott, S., and Smith, W. S., *Fuel Chem.*, 47, 81, 2002.

Dessler, A. E. & Parson, E. A., *Global Climate Change*, Cambridge Univ. Press, Cambridge, U.S.A., 2019.

Easterbrook, D., *Evidence-based Climate Science*, Science Direct Publ., New York, 2011.

El'gendy, N. S., and Speight, J. G., *Handbook of Refinery Desulfurization*, CRC Press, Boca Raton, FL, 2002.

Enns, R. G., *Nonlinear World*, Springer, New York, 2010.

Epstein, A., *Fossil Future*, Routledge, New York, 2022.

Ertl, G., *Encyclopedia of Catalysis*, vol.1, ed. J. T. Horvath, John Wiley & Sons, Hoboken, NJ, pp. 339–352, 2003.

Fanchi, J. R., and Fanchi, C. J., *Energy in the 21st Century*, World Scientific Publishing Co Inc., Hackensack, NJ, 2016.

Filho, W. L., *Handbook of Planetary Health*, Springer, New York, 2022.

Garrone, E., Fubini, B., Bonelli, Onida, C. O., *Chem. Phys.*, 513, 1, 1999.

Gates, B., *How to Avoid a Climate Disaster*, Vintage, New York, 2021 Kindle Edition.

GCCSI, *Global Storage Portfolio: A Global Assessment of the Geological CO_2 Storage Resource Potential*, Global CCS Institute, Melbourne, 2016 .

Havercroft, I., Marcory, R., and Stewart, R., *Carbon Capture & Storage*, 2nd ed., UK Publishing House, London, 2011.

Hinkov, I., Lamari, F.D., Langlois, P., Dickson, M., Chile's, C., and Pentchev, I., *J. Chem. Technol. Metall.*, 51, 600, 2016

IPCC (Intergovernmental Panel on Climate Change), *Climate Change*, Cambridge University Press, Cambridge, UK and New York, 1995.

IPCC (Intergovernmental Panel on Climate Change), *IPCC Special Report Carbon Dioxide Capture and Storage Summary for Policymakers*, Cambridge University Press, New York, 10, 2011a.

IPCC (Intergovernmental Panel on Climate Change), *Special Report on Carbon Dioxide Capture and Storage*, Cambridge University Press, New York, 2005.

IPCC (Intergovernmental Panel on Climate Control), Fourth Assessment Report, February 2007.

IPCC (Intergovernmental Panel on Climate Change), *Carbon Storage, Report for Policymakers*, Cambridge University Press, New York, 2011b.

Jose, P. D., *Astron, J.*, 70, 193, 1965.

Kamine, Y., and Chiangmai, A., *Handbook of Solar-Terrestrial Environment*, Springer, Berlin, 1965.

Kawamiya , M., Hajima, T., Tachiiri, K., Shingo Watanabe & Tokuta Yokohata. *Prog. Earth Planet. Sci.*, 7, Article number: 64, 2020.

Keith, David W., Geoffrey Holmes, David St. Angelo, and Kenton Heidel. 2018. *Joule* 2, 1573, 8, 2018.

Keller, J.U., Staudt, R., and Tomalla, M., *Berichte der Bunsengesellschaft fur physikalische Chemie*, 28, 96, 1992.

Kemp, L., Depledge, J. & Lenton, T. M., *Proc. Nat. Acad. Sci. USA*, 119, e2108146119, 2022.

Knudsen, *Ann. Physik*, 31, 205, 1917

Korotcenkov, G., *Handbook of Gas Sensor*, Springer, New York, 2013.

Kwok, D. Y., Li, D., and Neumann, A. W., *Langmuir*, 10, 1323, 1994.

Lacis, A. A., Schmidt, G. F., Rind, D, & Ruedy, R. A., *Science*, 330, 356, 2010.

Lamb, H. H., *Climate, History and the Modern World*, Routledge Publ., New York, 1995.

Leung, D. Y. C., Carmine, G., and Marato-Valor, M. M., *Renew. Sust. Energy Rev.*, 39, 426, 2014.

Li, J. R., Ma, Y., McCarthy, M. C., Sculley, J., Yu, J., Hae-Jeong, K., Balbuena, P. B., and Zhou, H.-C., *Coord. Chem. Rev.*, 255, 1791, 2011.

Lomborg, B. *Cool It*, Vintage Books, New York, 2007.

Lomborg, B., *False Alarm*, Vintage Books, New York, 2022

Lowell, S., Shields, J. E., Thomas, M. A., and Thommes, M., *Characterization of Porous Solids and Powders: Surface Area, Pore Size, and Density*, Springer, Dordrecht, NL,2006.

Lyman, W. J., Reehl, W. J., and Rosenblatt, D. H. (eds.), *Handbook of Chemical Property Estimation Methods*, American Chemical Society, Washington, DC, 1990.

Manahan, S. E., *Environmental Chemistry*, 11th ed., CRC Press, Boca Raton, FL, 2022.

Manuel, O. K., *Energy Environ.*, 20, 131, 2009.

McDonald, A. B., Thomas, M., Mason, J.A., and Kong, X., *Nature*, Cooperative Insertion of CO_2 in Diamine-Appended Metal-Organic Framework's, 519, 303, 2015.

Monson, P. A., *Microporous Mesoporous Mater.*, 47, 160, 2012.

Myers, A. L., *Langmuir*, 13, 4333, 1997.

Pessarakli, M., *Handbook of Plant and Crop Physiology*, 2nd ed, Marcell Dekker, Inc., New York, 2001.

Rabinovitch, J., and Govindjee, *Photosynthesist*, John Wiley & Sons, New York, 1969

Rabo, J. A., Elek, L.F., and Francis, J.N., *Stud. Surf. Sci. Catal.*, 490, 7, 1981.

Rackley, S. A., *Carbon Capture & Storage*, Butterwort Heinemann, Burlington, 2010.

Rols, J. L., Condoret, J. S., Fonade, C., and Goma, G., *Biotechnol. Bioeng.*, 427, 35, 1990.

Rosenzweig, C., Mutter, C., Contreras, M., (eds.), *Handbook of Climate Change And Agroecoscience*, World Science Publishing, London, 2021.

Rouqerol, J., Rouqerol, F., and Sing, K. S. W., *Adsorption by Powders and Porous Solids: Principles, Methodology and Applications*, Academic Press, New York, 1998.

Rowlinson, J. S. and Widom, B., *Molecular Theory of Capillarity*, Dover, London, 2003.

Sally, M. L., *Fundamental of Astrophysics*, Academic Press, New York, 1996.

Salty, M. L., *Physics of the Atmosphere and Climate*, Cambridge Univ. Press, Cambridge, 2012.

Silver, J., *Global Warming and Climate Change Demystified*, McGraw-Hill Professional, New York, 2008.

Sloan, E.D., *Clathrate Hydrates of Natural Gases*, 2nd ed., CRC Press, Boca Raton, 1998.

Stute, M.,Clement, A-., Lohman, G., Global Climate Models, *Proc. Natl. Acad. Sci., U.S.A.,* 10529, 98, 2001.

Sumida, K., Rogow, D. L., Mason, J. A., Mcdonald, T. M., Bloch, E. D., Herm, Z. R., Bae, T., and Long, J. R., *Chem. Rev.*, 112(2), 724, 2012.

Tanford, C., *The Hydrophobic Effect*, 2nd ed., John Wiley & Sons, New York, 1980.

Tovbin, Y. K., *The Molecular Theory of Adsorption in Porous Solids*, CRC Press, New York, 2017.

Tziperman, E., *Global Warming Science*, Princeton University Press, Princeton, NJ, 2022.

Yang, R. T., *Gas Separation by Adsorption Processes*, Butterworts Press, Boston, MA, 1987.

Index

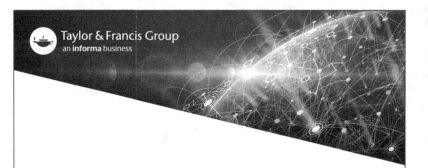

Printed in the United States
by Baker & Taylor Publisher Services